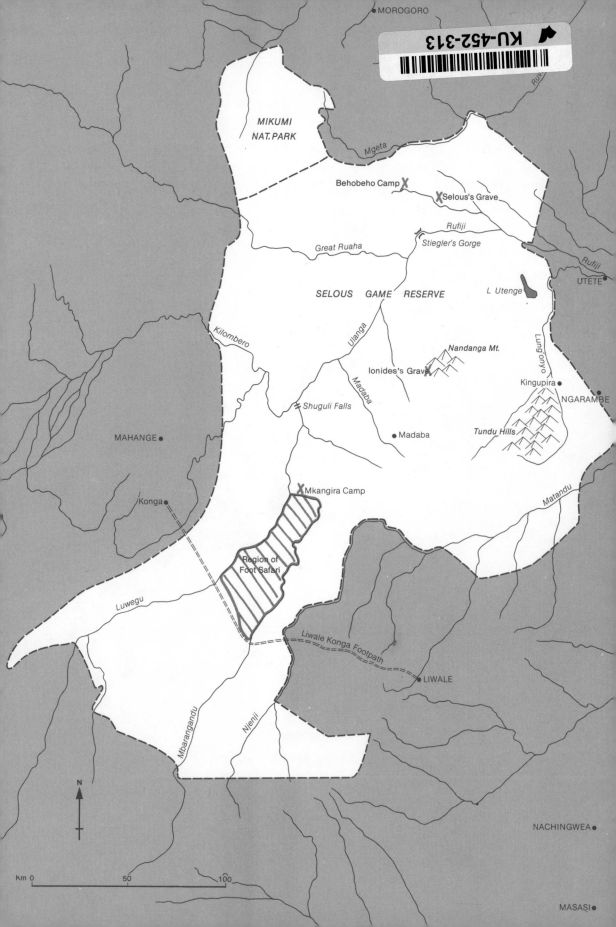

MOROGORO

KU-452-313

MIKUMI
NAT. PARK

Mgeta

Behobeho Camp ✘

✘ Selous's Grave

Rufiji

Great Ruaha

Stiegler's Gorge

Rufiji

UTETE

SELOUS GAME RESERVE

L Utenge

Kilombero

Ulanga

Nandanga Mt.

Ionides's Grave

Lung'onyo

Kingupira ●

NGARAMBE

Madaba

= Shuguli Falls

● Madaba

Tundu Hills

MAHANGE ●

Matandu

Konga ●

✘ Mkangira Camp

Region of
Foot Safari

Liwale Konga Footpath

Luwegu

● LIWALE

Mbarangandu

Njenji

N

NACHINGWEA ●

Km 0   50   100

MASASI ●

SAND RIVERS

Peter Matthiessen
SAND RIVERS

photographs by
Hugo van Lawick

AURUM PRESS

First printing

ISBN 0 906053 22 6

Printed in the United States of America
Colour separations by Latent Image, London
Designed by Terry Jones

For Eck and Donnie Eckhart
and
For John Owen

ACKNOWLEDGMENTS

Especially I wish to thank Tom Arnold MP for inviting me to join the Selous expedition, and Brian Nicholson, who put aside a healthy mistrust of writers in order to share his unique store of information and experience. All members of the expedition, African as well as European, made invaluable contributions that are identified within the text, and in addition, Hugo van Lawick made many helpful comments on the first draft of the manuscript.

Dr. Alan Rodgers of the University of Dar-es-Salaam provided a considerable amount of important information on the ecology of the Selous, and Drs. Iain Douglas-Hamilton, John S. Owen, George B. Schaller, Thomas Struhsaker, and David Western were all helpful on critical points.

P.M.

SAND RIVERS

I

Between two thoughts, in the month of August 1979, I discovered myself with mild surprise on the evening flight from London south to Dar-es-Salaam, in Tanzania: how often I had made this journey which in an airplane, at night especially, seemed as mysterious as a rite of passage. Imperceptibly during that night, in the silver time capsule of the long flight between continents, crossing the moon shadows of the Old World, imagining the wine-dark sea, the silent pyramids, the vast desert like a last barrier to the unknown, there came the sense that always comes on the way out to Africa, that the past had been left behind, that one was returning into the present, that one might emerge from this winged chrysalis with a new eye. And in a clear dawn of the next day, as the sun was born again from the Indian Ocean, lighting the dun thornbush and stone rivers of Kenya's Northern Frontier District, far below, a silence arose like memory from the turning earth and with it a promise and elation that I had rarely felt since childhood's morning.

The shining wing crossed the Equator at Mount Kenya, passing over into Tanzania as Kilimanjaro and Mount Meru rose like black islands from a sea of clouds. Tomorrow or the next day I would leave on a month's safari into the Selous Game Reserve, in southeastern Tanzania, said to be the greatest stronghold of large wild animals left on earth. Its area of 22,000 square miles makes it larger than Wales or Maryland; more to the point, it is the largest wildlife sanctuary on the continent,[1] almost three times as large as Kenya's huge Tsavo East and Tsavo West combined, almost four times as large as the great Serengeti National Park, in Tanzania. Yet of all the great parks and game reserves in East Africa, it remains the least accessible and the least known.

This "last safari into the last wilderness", as its sponsor, a young British Member of Parliament, described it, was to be led by a man I did not know, who had served as a warden of the Selous for more than twenty years and knew it better than anyone alive. If all went well, the safari would proceed to a remote point at the confluence of two great rivers in the far south of the Selous, where this man and I would cross the rivers and penetrate as far as possible on foot.

At the Dar-es-Salaam airport I was met by my friend Maria Eckhart, who was born at Moshi, at the foot of Kilimanjaro; she had left the United States two weeks ahead of me to visit her family at Njombe, in the Nyasa Highlands. Throughout the country, Maria told me, the economic situation was visibly more desperate than it had been when we were here in 1977. As we drove into the city it seemed to me that a look of apathy and discouragement was now apparent in the people's faces that had not been there two years ago, when there was even less in the shop windows. Because it was Sunday, when private driving is not permitted (no gasoline may be sold between Thursday evening and Monday morning), an emptiness wandered like blown litter through the town.

Besides the oil crisis, which has been especially harmful to poor countries, Tanzania was suffering from a year of failed crops and the attrition of its eight-month war to depose Field Marshal Idi Amin Dada of Uganda. According to a Maasai woman Maria had talked with at the airport, the people believed that Amin was now in the Sudan with three thousand of his troops, which was why so many Tanzanian soldiers had to stay on in Uganda; in any case, the country's foreign reserves were entirely depleted, and there was no money left to pay the troops. Many of those still in Uganda had taken up where Amin's soldiers had left off in the looting and exploiting of the local people; they completed the butchery, begun by the fleeing Ugandans, of the wild animals in the great Ruwenzori and Kabalega Falls National Parks.[2] Just three days earlier, a young wildlife biologist arriving in Nairobi reported[3] that Ugandans fleeing through the parks the previous March had killed a few hundred animals for food, but that the Tanzanians, who had also killed for food when they first turned up in the parks in April, had soon begun killing for profit, selling meat and skins as well as ivory. According to his "minimal estimate" more than six thousand hippo, five thousand antelope, and two thousand buffalo had already been slaughtered, mostly by machine gun fire and often from the backs of speeding trucks. Seventy lion and presumably many other creatures had also been killed; in his opinion the Ruwenzori National Park would be "dead in three weeks or less". The other Ugandan parks, beset by poachers since 1973, had fared no better.[4] Meanwhile, Kenya was struggling to recover from the epidemic poaching of the 1970s, which had threatened the existence of such splendid species as the black rhinoceros, reticulated giraffe, and Grevy zebra. In the light of all this sickening news, and of the chronic political instability that clouds the future of Africa, "the last safari into the last wilderness" did not seem such a fanciful description after all.

Tom Arnold, a pale-haired, friendly Englishman whom we met for the first time that evening, is a theatrical producer and politician; at thirty-two he was one of the youngest Members of Parliament, and had recently been appointed Parliamentary Private Secretary to the Secretary of State for Northern Ireland. Sometimes he came to Africa on official business, but more often he made the journey for its own sake. "On my first visit," he told us over a drink in the bar of the hotel, "I was enchanted by Africa; that's the only word." In the previous few years he had made seven trips, and these days he spent as much of his free time here as possible, usually in the company of the former warden of the Selous.

The enchantment had set in during the winter of 1976, when Arnold took the standard tour of the southern Kenya–northern Tanzania game parks, then extended his trip to include a flying visit to the Kenya coast and to the northern game park called Samburu. The charter pilot who flew him to Samburu was a man named Brian Nicholson, who turned up there again the following day. "You think *this* is impressive, do you?" he said to Arnold, and went on to describe the Selous Game Reserve, a little-known wilderness in southeastern Tanzania where he had served as game warden for many years. "I couldn't believe a place like that could still exist," Arnold said. "Like most people I had never heard of it, and I wanted to see it for myself." That same year, Arnold made four trips out to Africa, and on the last of these, in early autumn, he hired Nicholson to fly him down to the Selous. The former warden was upset by what he saw as the utter abandonment of the system of dry-season tracks, air strips, and patrol posts that he had established there between 1950 and 1973, but what Arnold saw was what Nicholson had promised, the largest and wildest game sanctuary in all East Africa. "There was an enigma about it, a mystery – that vast wild place that scarcely anyone knew."

Before long, Arnold had hit upon his idea of an ultimate safari. As he discovered when he sought to inform himself about the place, almost nothing had been written about the Selous: because it was not a park but a game reserve, with no facilities for visitors, this remote and inaccessible region had been all but ignored in the great pilgrimage to the wildlife of East Africa that had seized the imagination of the Western world in the 1960s and 1970s. Baron Hugo van Lawick and myself were asked to do the book that would help to underwrite his expedition, a book about a Selous safari led by the man who knew it best, and a number of others were invited as guests, including Maria Eckhart, the Nicholson family, and David Paterson, a young Hong Kong businessman who would later join

Tom Arnold as an expedition sponsor. Meanwhile, Nicholson had contacted a fellow pilot named Richard Bonham, who was a director of Nomad Safaris operating from Mombasa, a new company that avoided the usual tourist routes in favor of real expeditions, mostly to the Northern Frontier District and the Tana River. Kenya safari companies were no longer allowed to operate in Tanzania, on account of the bitterness which in 1977 had led to the closing of the border; but since, in the absence of supplies, no Tanzania company was equipped to handle the caravanserai that Arnold had in mind, Nicholson persuaded friends in the Tanzanian government to waive the rule, with the good argument that international publicity for the Selous might help to save it. By mid-August the Nomad caravan – a truck and several Land Rovers – had crossed the border, and even now was headed south toward our first base camp in the Selous, accompanied by Hugo van Lawick, who was bringing his own Land Rover from the Serengeti.

Even with cooperation, things had not gone smoothly. In recent months, Bonham had had to make several trips to Dar-es-Salaam to confirm frontier clearances and permits, and on the last of these, in early August, he was detained at the airport for ten hours because officials there had mislaid his permit to fly into the country. That afternoon Tom Arnold and the Nicholsons, arriving from Nairobi by charter plane, were refused permission to enter the country for three hours. Tom Arnold told us that he had taken the precaution of escorting Nicholson out of the Immigration Office. "He takes this sort of thing very badly," Arnold said. "Remembers how things used to be, you know." He laughed loudly in amusement and affection. Finally the British High Commissioner's assistant, called out on a Sunday, came to the airport and discovered the necessary clearance among the papers on the official's desk.

A tall, thin man with a closed face appeared, leading his women. Arnold stood up to introduce us to Brian Nicholson. Somehow I had come by the notion that this legendary warden was "a bit difficult" – that is, grumpy and suspicious of "outsiders", and writers especially. In any case, we were wary of each other, and because neither of us have much taste for small talk, the working relationship that was crucial to the success of the safari started out with an interminable pause. Finally I asked Nicholson why he thought the immigration official at the airport had behaved so arbitrarily when the clearance was sitting right there on his desk. "Right hand doesn't know what the left hand is doing," said the Warden shortly. Was it possible that this was "official" harassment of visitors from Nairobi, due to the bad feeling between Kenya and Tanzania? "They're not that efficient," Nicholson said.

Unlike his wife, Melva, and his pretty daughter Sandra, who were very friendly from the start and tried to ease things, Nicholson seemed cross, stiff, and uneasy, and scarcely ever looked me in the face that whole first evening. In profile, which I had ample opportunity to study, the Warden was rather a handsome man, much younger than I had expected, with weather lines and good, high color and taffy-colored hair in a military cut, shaved high at the back and well above the ears. Head on, however, his down-drawn face appeared slightly askew. It wasn't, of course; the imbalance lay in his expression. His full lower lip had a slight twist, sardonic, faintly truculent, and a bald cast to his pale blue eyes gave an air of coldness that his brusque manner did little to offset.

At supper, overlooking the night harbor, Maria and Tom managed to draw the Warden out. Maria soon discovered that Brian, who was forty-nine, had attended the Prince of Wales School in Nairobi (known as the "Cabbage Patch", she reminded him, due to the "smelly boys", who referred to her own Kenya High School for Girls as the "heifer boma", or young cow pen). One of his schoolmates there was Bill Woodley, perhaps the best known of all East African game wardens, who has remained a lifelong friend, and another was Maria's older brother, Peter Eckhart, known to his fellow scholars in those grand old colonial days as Njoroge or "Nigger" because of his black curly hair. "Oh yes, he was in the form behind me, a little red-faced boy," Nicholson recalled, reflecting mildly after a moment, "I probably made him cry." Maria stared at him, startled. "You probably did!" she exclaimed at last. "Oh, I know his type," she told me later. "Typical East African redneck, a real Boer!" But like myself, she sensed already that Brian Nicholson was a good deal more complex than the anachronism he seemed to wish to represent. He was very much his own man, but aggressively so, emphasizing his own self-sufficiency, not always obliquely, in almost everything he said.

"That gentleman we were talking to –" Tom Arnold was saying, and Nicholson said in honest puzzlement, "*What* gentleman? Oh! you mean the *African*."

The disparity between these two was disconcerting. Tom Arnold had an aristocratic manner to go with his obvious intelligence and education, and he was as politic and well-spoken as the other was abrupt; he was prone to big laughter rather than guarded, as well-fleshed and fond of wine as Brian Nicholson was lean and abstemious, and as plainly a man of the indoors as his new friend was ruddied by the sun. Consciously or not, Brian Nicholson was playing the role of the crusty old-style bush type who had been described to me, and that first evening, the month ahead seemed a very long time indeed.

For all his stiffness, Nicholson was at pains to be obliging, and presently suggested that I come up to his room after supper to have a look at his maps of the Selous. Apparently, he had made a decision to cooperate in Tom Arnold's safari and to work with this bloody American writer, whatever the cost. I noticed quickly that the going was a lot easier when we were away from other people, but at the end of the first evening I still did not know if his closed expression derived from a fundamental suspicion of all writers or merely a simple aversion to myself. Tom

Arnold had previously mentioned to me that Nicholson had read a book I had written on East Africa: "Seems to know something about Africa," he had muttered, in what Tom diplomatically assured me was high praise. Yet his apparent views on race and politics did not seem to accord with those expressed in *The Tree Where Man Was Born*. On our way out to the University of Dar-es-Salaam the following day to talk with one of Nicholson's old colleagues, an expert on the ecology of the Selous, he did not hide his disapproval of the way that Africans were managing here in the city; of the epidemic idleness and inefficiency, brought about, in his opinion, by a "system that maintained people in jobs because they were black, not because they were qualified"; of the weeds and chickens in what had been the gardens of his house out at Oyster Bay; of the chronic failures of the telephone, electricity, and even hot water for his tea in what was supposed to be the best hotel in Tanzania (the news that the cash register had broken down made him laugh, the first laugh that he ever emitted in my presence, and not a heart-warming sound in any way). I decided not to ask him his opinion of the notices the hotel had posted in the rooms, under the heading USEFUL HINTS, which cautioned the traveler to avoid the waterfront and harbor area, especially after dark, and made the following suggestions:

> Be wary about "hard-luck" stories from tricksters hanging around, with their smooth talk and friendly persuasion.
> If at all a taxi is required, use one from a reputable firm.
> Ornaments are very tempting for local snatchers.

For those who knew which of the motley taxis might represent a reputable firm, such hints *were* useful, unlike the notice seen by Hugo van Lawick, posted by the local commissioner next to the telephone in the Kigoma airport: "It has come to my attention that use of this telephone is absolutely disgusting."

II

On 21 August we flew inland over the cloud shadows and small settlements of the coastal bush, entering the northeast corner of the Selous at a point not more than seventy miles southwest of the capital. The plain beneath had been turned hard black by recent burning, and I asked Brian, seated in front next to the pilot, Godfrey Mwela, if he thought the fires had been set by Game Department people to generate new growth. He shook his head. "Poachers, I should think. Always had a problem with them in the north because here the villages come up so close to the boundaries. I've seen snares strung through here for almost fifteen miles, and dead animals rotting all along the way."

Off to the west rose the hills of the central African plateau, and soon the plane was crossing the new swamps created between 1971 and 1974 by the Rufiji River, which had left its ancient course as sandbanks built up in the main channel and overflowed onto this eastward plain below Stiegler's Gorge. It was now mid-afternoon, and large groups of elephant and buffalo were moving peacefully toward the shining water. Headless palms turned like strange sentinels under the shadow of the plane; most of the trees had been killed by the villagers, who lop off the crowns, then convert the rising sap into "palm wine". Here and there stood the delicate tall stalk of a giraffe, the first and last giraffe we were to see in the Selous.

For reasons not well understood, the giraffe of eastern Tanzania does not occur south of the Rufiji although the same species reappears beyond the Zambezi River, more than six hundred miles further south; and the white-bearded gnu is replaced below the Rufiji by a distinct morphological race, the Nyasa wildebeest, which lacks the white beard but has a striking white chevron on the forehead. It has been suggested[1] that the course of the Rufiji represents an ancient fault line, connected to the Rift, that was once too deep and dangerous to cross and served as a true geographical barrier which permitted the separation of the wildebeest into two subspecies. But this theory of the "Rufiji barrier" does not explain why the giraffe and a number of smaller creatures have failed to extend their range now that those hypothetical deep canyons have silted in, nor why the two races of wildebeest should remain distinct, with no trace of intermediate forms, when all that separates them for half the year are broad sand bars and the shallow channel of a river that waterbuck as well as elephant and buffalo cross without difficulty.

Under the hills the plane banked toward the north, following the small Behobeho tributary. In an open plain of game trails and scattered trees stood the white concrete rectangle, inset with a marble plaque, that marks the grave of Frederick Courtenay Selous, naturalist, elephant hunter, and explorer, after whom the Reserve was named in 1922. Selous, who was once "white hunter" for Theodore Roosevelt, was a captain in the 25th Royal Fusiliers, "the Legion of Frontiersmen" – "bush types, mostly," according to Nicholson – part of a far superior British force tied up for years by guerrilla troops under the command of the remarkable Count von Lettow-Vorbeck, in a useless battle to control the *bundu* or bush of this wild country. The elusive German and Selous, far out on the uttermost frontier of the Great War, had the time and perspective to deal with each other as gentlemen, and von Lettow, in his memoirs, makes the claim that he once had Selous in his rifle sights but let him go. A common soldier, it appears, was less impeccable, and Selous died of a wound received at Zogoware, not far from this place, in the course of the British advance to the Rufiji in 1917.

The Germans had established a game reserve in this region as early as 1905, and the several reserves established by 1912 were known collectively to the Africans as Shamba ya Bibi, or "Wife's Land" – the Kaiser Wilhelm, on a romantic impulse, having made it a huge and shaggy present to his wife. No Germans inhabited the region (their forts were nearer the coast, at Utete and Liwale) although in 1916–17 they set up gun emplacements along the Rufiji and the Behobeho rivers, where old horseshoes, cartridge cases, and the like may still be found. They also hauled an enormous steam engine all the way inland from Kilwa, using hundreds of unfortunate Africans as draft animals, to grind millet to make bread for their black soldiery; the abandoned steam engine still serves to commemorate the Second Reich in the Selous Game Reserve, as the collected reserves were named in 1922 by the Game Department established that year by the British. In those days the Reserve was confined to a tract of approximately one thousand square miles, and not until a decade later, with the advent of an inspired young game ranger named C. J. P. Ionides, did it begin to assume the imposing shape that it has today.

Ionides – or "Iodine", as he came to be known throughout East Africa – has been called "the father of the Selous" by no less an authority than Brian Nicholson. A former British Army officer turned ivory hunter, he was briefly a white hunter working out of Arusha in 1930, then joined the Game Department in southeastern Tanganyika in 1933. Although he continued in his avid hunting, collecting rare species as far away as the Sudan and Abyssinia, Ionides was a precocious conservationist. Very early in his career he began to envisage a great and self-perpetuating bastion of African wilderness, a complete ecosystem (as it is known today) where animals might wander in merciful ignorance of human beings. During a tenure of more than twenty years, he devoted his formidable energies to advancing this concept from a political point of view, exploring the region to map out some sensible boundaries and discouraging the activities of its scattered tribesmen, wherever possible by removing them entirely.

As game ranger Ionides was responsible for reducing the trouble-some elephant populations of the Kilwa and Liwale districts, which were causing serious damage to the isolated shambas and small villages. With his limited staff, effective control of the wide-ranging animals was impossible, and eventually he got permission from the colonial adminis-tration to deny crop protection except to those Africans who moved to certain designated settlements. The British authorities agreed to this plan because it simplified tax collection and other administrative duties, but their support of it was only intermittent, and Ionides, who had already trekked extensively over the country, suggested in 1935 that *all* of the western Liwale region be made a game reserve in which human settlement would be discouraged, thereby returning this wilderness, with its poor soils and abundant tsetse fly, to the wild animals for which it was best suited. In 1936 this proposal was accepted, mostly because of an outbreak of sleeping sickness in the affected area: consolidating the isolated inhabitants into communities with medical facilities, in which bush-clearing and eradication of wild animals might control "fly", made excellent sense. In the next three years, as the disease persisted, Ionides kept right on walking, mapping out the proposed extensions of a vast new elephant reserve in western Liwale that would extend northward to the existing Selous Game Reserve. But when these new boundaries were laid out, in 1940, a small number of people remained inside them, and the colonial administration balked at the actual eviction of the inhabitants. Ionides persisted in his policy of denying all protection to outlying shambas in western Liwale, and after three more years of severe damage the last of the Ngindo tribesmen in the southern region gave up and moved to settlements outside the Reserve. Once they were gone the territory was declared an elephant reserve, so that they were legally prohibited from moving back. As it turned out, all these settlements were located so far outside the boundaries that to this day, in the southern region, the country all around is virtually as uninhabited as the Selous itself.

Meanwhile the epidemic had spread north and west, leading finally to a forced evacuation of the Africans that was complete by 1947. The main outlines of the modern Selous were established by law in 1951, although important additions were subsequently made by Brian Nicholson. From start to finish, as Nicholson has written, the expansion of the Selous was resisted by the administration. "This was largely due to the fact that the people evacuated during 1946 and 1947 did not believe that they were moved because of sleeping sickness, and since most of the area became Game Reserve, the idea grew that the real reason was to facilitate the creation of this Reserve . . . To the present day the Game Division is treated with considerable reserve by the people in Kilwa, Liwale, Rufiji and other Districts bordering on the Selous."

As in the case of so many great wildlife sanctuaries of Africa, the nominal factor in the final creation of the Selous Game Reserve was the tsetse fly. It has recently been argued[2] that tsetse epidemics causing trypanosomiasis or "sleeping sickness" in human beings are not natural events but are brought about by the white man's interference, specifically the disruption caused by colonial policies of stock raising and land management, which upset the delicate, unknown balances between pastoral Africans and the domestic and wild animals. But the bush-loving Ngindo in the Selous had no domestic animals and were apparently content to live with tsetse; and in any case the epidemic died out almost as rapidly as it had appeared, to judge from the fact that in the 1950s and 1960s not one of the hundreds of Game Department staff who inhabited the Selous for long periods was ever to come down with the dread disease. When I asked him about this, Nicholson grinned. "That sleeping sickness wasn't quite so serious as old Iodine had the authorities believe. But he saw the great chance to accomplish what he knew was best, and he just took it." He shook his head over the boldness of the man, and I did too. Before Ionides was finished, Shamba ya Bibi had been enlarged nearly twenty times from the original tract of approximately one thousand square miles, and its dramatic expansion was the direct result of the organized depopulation of vast areas of southeast Tanzania.

A number of people I have come across in Africa had been friendly with this extraordinary man, who died in 1967. Several books based on interviews with Ionides were put together in his lifetime, since he was exceptionally colorful as well as single-minded; and he himself wrote interesting papers for the *Tanganyika Notes and Records*. Yet one is led by his own statements to suspect that, like most so-called "eccentrics", Ionides was more presentable in books than he was in person, at least in any acquaintanceship at close quarters. Born in 1901 of a rich Greek family well-established in southern England, he found himself snubbed and isolated at the snobbish Rugby School because of his "foreign" origins and appearance and his sickliness, which were held accountable, no doubt, for his equally suspicious interest in wild creatures; eventually he was driven out of Rugby under unjust circumstances that seem to have embittered him for the rest of his life. Even at school his physical courage and stoicism when beaten were legendary, yet in order to prove himself he won a commission in the Army (where his nickname was "Greek"). But from the start he was a non-conformist, and very early in a promising career he retired from the Army, taking up solitary hunting expeditions that eventually led to his self-isolation in this most remote region of the East African bush.

From boyhood, Ionides's hero – by all accounts, the only one he ever had besides Genghis Khan – was another old Rugby boy of an earlier era, an inspired amateur naturalist and "the greatest hunter of them all",[3] Frederick Courtenay Selous, whose book *A Hunter's Wanderings in Africa* had much impressed him, and whose death in 1917 he considered a "personal loss". Though the flinty Ionides would never have confessed to such sentimentality, his identification with Selous might well have affected his efforts on behalf of the Reserve.

On the ridge above Selous's grave stood Behobeho Camp, once a hunting lodge owned and operated by Ker & Downey, who sold it to the Oyster Bay Hotel in Dar-es-Salaam in 1970. There are two other tented camps for tourists in this northeastern corner of the Selous, on the Rufiji flood plain, and there is also a temporary settlement on the Rufiji River inhabited by Norwegians who are building a bridge dam over the river at the upper end of Stiegler's Gorge. Designed to be finished about 1990, the dam will create a lake that, in the rainy season, will be about 650 square miles in extent. In itself, the lake will do little harm to wildlife, since unlike these great animal plains the region of the upper Rufiji that will be flooded is relatively barren; but the predicted influx of fishing people, and the shifting agriculture that is bound to occur along the access roads, may make it difficult to enforce the present status of what is today the largest tract in Africa in which no human being has rights of settlement or even entry. While the dam will regulate the flow of water to the rich agricultural flood plains below Stiegler's Gorge, it may also threaten the estuarine fishery of this largest river in East Africa and perhaps even the offshore fishery for prawns, and is therefore regarded with misgiving by ecologists. Meanwhile, the new road across the dam may increase the pressure for development by opening up the eastern part of the Selous.

Godfrey Mwela turned his plane east again, following the rocky chasm of brown swift water for perhaps five miles until it opened out on the coastal plane below. There the plane turned south, heading out across a trackless wilderness. From this place for hundreds of miles to the south and west there are no habitations or facilities of any kind except the Game Department station at Kingupira, about eighty miles away, where we were to join the rest of the safari.

Because it is near all four habitat types of the Selous – the alluvial hardpan plain of the eastern border, the riverain forest, the *Terminalia spinosa* woodland, and the *miombo* or "karoo" that covers three-quarters of the Reserve – Kingupira was chosen for the site of the Miombo Research Center, set up in 1969 under Nicholson's direction by a young ecologist named Alan Rodgers. Rodgers, whom we had visited at the University of Dar-es-Salaam, was a bluff, husky, generous man who dispensed beer as well as his own documents, and also a fascinating discourse on the ecology of the Selous, where he had spent ten years; according to Nicholson, whose admiration he reciprocates, he is the greatest living authority on the ecology of the *miombo*, the vast wilderness of "dry forest" or savanna woodland which extends almost from coast to coast for sixteen hundred miles across the waist of Africa. *Miombo* is very similar to the so-called Guinea savanna of West Africa, but whereas Guinea savanna occurs in a narrow belt between the expanding Sahara Desert and the tropical rain forest, the *miombo* in places extends north and south for well over a thousand miles.

Alan Rodgers, who served as Game Research Officer in the Selous from 1966 to 1976, has no lack of arguments to support his high opinion of what he has referred to as "the sleeping wilderness".[4] With dependable rainfall, great rivers, and innumerable springs and seepage points, there is abundant ground water and pasturage in the Selous Game Reserve, and since its boundaries enclose three distinct ecological units (the Serengeti cannot claim even one), there is no need for its 750,000 large animals to move out into unprotected areas, or "migrate" to water or dry-season pasture, or herd up anywhere at all in vulnerable or destructive concentrations. Most of the reserve is between 1,000 and 2,000 feet above sea level in humid country not far inland from the coast, and the vast *miombo* tracts are isolated from one another by the barriers of the Rufiji River and its tributaries, especially the Ruaha, Kilombero, and Luwegu. The Kilombero delta and the open country between the great rivers in the south are very beautiful, but the only scenic region accessible to visitors is the one where the tourist camps are already located, less than one hundred miles from Dar-es-Salaam. That open Behobeho country, in Rodgers's view, is "some of the most magnificent wildlife country in East Africa", while the rest of the Reserve is decidedly unsuitable even for the local people, despite chronic demands for fishing, grazing, farming, and timber rights, mostly from the Mahenge district, which are bound to increase with the completion of the dam at Stiegler's Gorge. Apart from the problem of tsetse, the soil almost everywhere is leached out and eroded, and the *miombo* woods are of little use except for the gathering of wild honey: in fact, the management of wildlife is almost certainly the most productive use to which this land could be put. The gathering of people into settlements that could be efficiently administered by the colonial government has been continued in independent Tanzania by the *ujamaa* villages, where the people are very similarly administered in the name of socialism: except in certain places in the north, the scattered human populations along the boundaries have actually decreased as a result of *ujamaa*, and this removal of human pressure, together with the difficulty of access, the abundance of game outside the boundaries, and the check on human predation by the *ujamaa* administration, have all served to reduce poaching, which since 1960 has been negligible compared to the activities in other wildlife sanctuaries of East Africa. As a result, the Selous can claim East Africa's largest populations of many creatures, including elephant, rhino, buffalo, hippo, and crocodile (lion and leopard, too, according to Nicholson), as well as more than three hundred and fifty species of birds and two thousand species of vascular plants. According to Rodgers, "It is unique in its size, its state of naturalness, and its variety of genetic and ecological resources."

On all sides as we flew south the open woodland was broken by scattered pans of glinting water; in the southern distance, clouds of smoke rose all along the low horizon. "Whole country's going up in smoke," Nicholson muttered, not without a certain grim satisfaction; he believed strongly in the use of fire for eliminating the dry grass and encouraging new growth to support more animals, but having little faith in the present Game Department, he suspected that most of these fires were set by local poachers, for whom hunting was easier when the landscape was burnt.

Godfrey Mwela dodged two vultures that suddenly came sweeping down on his propellers, and seated behind him and Nicholson I could see the wrinkles of pleasure at the Warden's eyes as he laughed and joked with his fellow pilot. ("Best pilot in Tanzania, Godfrey is; helped to train him myself.") Nicholson turned to me repeatedly to point out the flat-topped terminalia trees that dominate this eastern woodland, the arrangement of termite hills around the pans, the tamarind and mahogany trees that the hills support, Nandanga Mountain in the fire haze off to the west – "That's where Iodine is buried, Peter!" he said, calling me by my name for the first time. Brian caught me by surprise; in the excitement of coming home to the Selous, he had betrayed an unabashed enthusiasm, not only to me but to Godfrey Mwela, whom he was addressing not just civilly but as a friend.

Later, as if to readjust a mask, he made an unpleasant colonial joke, expressing surprise that one of Rick Bonham's Kenya staff had been with us on the plane. "Must have been back there in the dark," Nicholson said.

"He didn't smile, so I never knew he was there." He laughed slyly at our cold expression. "That's the kind of joke my brothers used to make," Maria said later, "but at least they outgrew it." I was beginning to suspect that Nicholson had outgrown it, too, that being obstreperous was just another way of saying that he didn't give a damn for the world's opinion. (Or perhaps, as a friend who was raised in Dar once said to me, "As white East Africans, we feel we *have* to talk that way. I don't know why.")

Rain came and went. Near the Ngindo village of Ngarambe, the plane crossed over the Lung'onyo River that forms this part of the eastern boundary, then swung back again to make a good landing on the strip at Kingupira, startling two wart hogs out of the long grass.

We had scarcely unloaded when two Land Rovers turned up. The first people to jump out were two old game scouts, Saidi Mwembesi and Bakiri Mnungu, who were so delighted to see "Bwana Niki" that they clung to his hands throughout the protracted greetings that African courtesy demands: both used the respectful Arabic salutation *Shikamu*, meaning "I kiss your feet". Bakiri could not get over how Sandra Nicholson had grown, how time had passed. He slapped his head, hooting with laughter: "See? I'm losing my hair!" Bakiri is a local man and still works with the Game Department, but Saidi, a very tall Ngoni Zulu in a Muslim cap, had quit the Department at the same time as his Bwana and gone away to live in Dar-es-Salaam, where Rick Bonham had picked him up on the way through. ("He's a great old chap," Rick told me later, describing how Saidi had replied when the senior warden asked him if he still remembered the Selous well enough to guide Mr. Bonham down to Kingupira. "Remember the Selous?" Saidi had snorted, brandishing both hands. "The place was built with these hands! *Mkono yangu!*")

Karen Ross, who was in charge of the safari kitchen, and Robin Pope, a young assistant from Zambia, were also there to meet us, and soon Hugo van Lawick came in his own Land Rover, which is specially fitted out for photography; van Lawick had just observed five pups at a wild-dog den, less than a mile from our camp.

A comfortable cluster of green tents awaited us in a pleasant copse of palm and tamarind along the edge of the Kingupira Forest. We were greeted by Philip Nicholson, aged nineteen, a likeable blond boy and fanatical fisherman who was making a last safari with his family before going off to seek his fortune in Australia. The only person not yet among us was David Paterson, who was to fly in the following week with Rick Bonham's supply plane to the airstrip at Madaba, seventy miles to the southwest.

This camp by the forest, a few miles away from the present Game Department field headquarters and the near-moribund Miombo Research Center, had been one of Brian's first safari camps and also the site of an early game post. Among the courtesies extended to our expedition by the Game Department was a resident's permit to shoot impala, buffalo, and guinea fowl, to help feed its seventeen black and white participants, and after a fine first supper of impala and red wine, we sat around a campfire under the stars. Though at pains not to show it – he was already grumping about the scarcity of his beloved elephants – Brian Nicholson was very happy to be back. For the first time since I had met him, he did most of the talking, describing how he had first come to the Selous.

At the age of sixteen, Brian Nicholson abandoned his formal schooling in Nairobi and went to work for the noted animal collector Carr Hartley, who paid him "one hundred shillings a month and *posho* (rations)" and gave him his first lessons in dealing with large wild animals. The next year he signed on with a professional hunter named Geoffrey Lawrence-Brown and made thirteen or fourteen safaris as an apprentice hunter in order to qualify for his professional hunter's license. But being a "white hunter" did not interest him. He had always wanted to be a game warden, which in those days, he said, meant living in the bush and hunting and shooting to protect the shambas of the local people from the depredations of wild animals, especially elephant and lion. No such jobs were available in Kenya to a youth of his experience, and in 1949, at the age of nineteen, he signed on as a "temporary assistant elephant control officer" assigned to the region of the ill-fated Ground-Nut Scheme, which had its main headquarters at Nachingwea, in southeastern Tanganyika. "My qualifications were virtually non-existent," Brian said, "but nobody else wanted the job."

In its first years, the Ground-Nut Scheme was a threat to the southern Selous, which the planners thought might prove suitable for agriculture; in the absence of roads to Dar-es-Salaam, they planned to export 440,000 tons of ground-nuts annually through the new deep-water port being specially developed at Mtwara (where Maria's father was asked to set up a hospital, and where her sister Patricia was the first white baby to be born). But the Scheme collapsed under accumulated folly long before its grandiose ambitions could be implemented; among its many serious miscalculations was the failure to realize that by harvest time the ground-nuts planted in the soft mud of the rains would be locked under the hardpan of the dry season, and would have to be chipped out one by one. A little late, the planners asked themselves why this region was so thinly populated in the first place.

Nicholson's supervisor was Constantine John Philip Ionides, then Senior Game Warden of southeastern Tanganyika and already a notable collector of rare animals and poisonous snakes; it was he who reported, for example, that the green mamba and the Gabon viper, at that time considered to be largely West African in distribution, were in fact very common on the Makonde Plateau, where Tanganyika bordered Mozambique. (Maria's father, who remembers Ionides with fondness, once told me that the hospital verandahs at Mtwara were a favorite place for collecting cobras.) Ionides and young Nicholson took to each other straight away, and the next year Brian transferred to Ionides's headquarters at Liwale near the southeastern boundary of the Selous and began the long series of foot safaris that were to acquaint him with most of the southern country.

By the time Nicholson appeared, in 1950, Ionides considered that his great work of creation had been done; he was devoting more and more time to the hunting and collecting that had become his passions, and increasingly so as he realized that this new young assistant in elephant control whom the Africans called Bwana Kijana (the Young Bwana) was capable of taking over most of his duties and carrying out the final steps in his master plan. At one point in 1951, when Ionides was off in the Sudan hunting for addax, Bwana Kijana was called in by the Provincial Commissioner at Lindi in regard to a Game Department request to incorporate this Lung'onyo River region into the Selous. The P.C. stepped over to a wall map and slowly traced the expanding outline of the Selous with his thumb, then said coldly to Nicholson, "You people are sterilizing this whole area." Remembering this, Nicholson remarked, "That man could never look you straight in the eye. Don't know where they found such people – hadn't a clue about Africa. All they thought about was giving these local Africans just what they wanted, whether it was good for them or not. No thought for the animals at all!"

In 1954 Ionides went off on leave and, except to collect his things, never came back; he formally retired from the Game Department in order to give full time to collecting uncommon creatures on commission for various clients, including the Coryndon Museum in Nairobi. Since renamed the National Museum, it still displays Ionides's gorilla group, bongo, and addax – pursued on camel back – and an assortment of other creatures, including the black mamba that, in 1942, crawled over his bare legs in the dark while he was seated in an outdoor privy in Liwale. Ionides later credited this creature with inspiring the snake collecting avocation that eventually displaced the hunting of rare animals as the great passion of his life. Of the local Provincial Commissioner – perhaps the very one disliked by Nicholson – Ionides once remarked, "He wasn't an attractive character: he didn't like snakes and he had beady eyes and damp hands." Ionides was fond of saying that he found human beings the least interesting of all animals, which may have accounted for his reaction when in the dark a green mamba fell out of the thatch roof of a local hut on to a group of sleeping Africans and bit eight of them fatally before escaping from the panic that ensued. Ionides, who succeeded in capturing this snake, was outraged: "If I hadn't been in that area, they'd have pursued it relentlessly and beaten it to death with sticks! Hooligans, insensitive dolts, thoroughgoing bastards!"

Like many another wounded by snobbery early in life, Ionides became a snob himself, "a Royalist and imperialist of deepest-dyed hue", as he himself described it, a "dinosaur", to use the word of one of his colonial contemporaries, at least in all matters having to do with Africa and Africans. Before he acquired the name Bwana Nyoka, or Snake Man, Ionides was known as Bwana Mparangozi, or He Who Takes the Hide Off Them, a name awarded for his unstinting use of the *kiboko*, or hippo-hide whip; on one occasion he prescribed flogging for all sixty adult males of a certain village. Ionides later acknowledged that he might have been a bit too free with the *kiboko*. However, he said, "Flogging is a method that is simple and effective and very widely understood. . . . In a primitive country you use primitive methods. For twenty-five years [c. 1935-60] this country's had a rule of fish-flabby hands in velvet gloves, and all it's done is to make third-rate Europeans out of a race of potentially first-rate Africans." In the same vein he remarked a few years later, "These people have been largely emasculated, that's what it amounts to. Their splendid virtues have been driven out of them; it isn't their fault if we've turned them into a rather second-rate lot."

Quite apart from anything else, a certain nostalgia is apparent here: the man of "Old Africa" has always been more attractive to the white man then the "new African" who presumes to compete with him. Thus the unregenerate Ionides, speaking at the time of Independence, could still refer to "our little black brothers, who haven't yet even developed a brain to think straight with, and certainly haven't contributed a single idea to the present sum-total of the civilization of mankind."[5] One has to wonder about the effect of such an attitude on a very young, half-educated Kenyan from the White Highlands, which were already being claimed by the restless Kikuyu who would revolt in the Mau-Mau rebellion a few years later.

On one of the first of his long safaris Nicholson visited the Lung'onyo River, making his camp where we now were by the Kingupira Forest. There had been a report of heavy poaching, and everywhere he found hunters' blinds set up at the water pans and thorn fences that guided the game into set snares. In those days, poaching was a local enterprise, mostly for meat, which because of the tsetse fly and the virtual absence of livestock was in heavy demand. In a nearby Ngindo settlement he discovered some large stacks of hides, and sat himself atop a termite hill while his game scouts burned down the thatch village and placed its men under arrest. At the people's request, he spared the *ngokwes* (stilt storage huts) until the grain could be removed, and the next morning, several hours away along his route, he sent a few men back to finish off the job, under the leadership of a huge Ngoni Zulu named Nonga Pelekamoyo. ("The Ngoni like to give themselves resounding names; 'Pelekamoyo' means 'take your heart'.") Doubtless inspired by the memory of the great days when the Zulu armies from the south swept through this country, scattering the fearful Ngindo into the bush, Nonga

Take-Your-Heart put the torch to every community that he passed, an estimated ninety huts in all, an outrage for which, the following year, Bwana Kijana was summoned to Game Department headquarters at Arusha, and forced to stay there for approximately six months. "This old man who met us at the airstrip" – Brian pointed at Saidi Mwembesi – "he's the nephew of old Nonga, and he has a son who's in the Game Department now. Those people are a Game Department family."

In 1955, when Brian took his first leave from the Game Department, he was no longer Bwana Kijana but a full-fledged Bwana Nyama, or "Mister Game", as the Africans called the Game Department wardens (or "Mister Meat", Brian said wryly, choosing the other meaning of *nyama*), and so absorbed in his job that he did not want to take leave at all. "In those days," he says, "they used to make us take overseas leave, and so I decided to visit the U.K." At the Overseas Club in London, where he went for want of any destination, he met a pretty Australian girl named Melva Peal. Miss Peal came out to Africa in August of that year, and despite the warnings of Ionides, who had avoided women all his life, and who assured Nicholson that domesticity would never mix with a life in the bush, they were married in January 1956, proceeding immediately to Nachingwea. This erstwhile hub of the Ground-Nut Scheme, with its grandiose avenues and city planning, "was a dead city by the time I got there," Melva said. "There were hardly any Europeans left. I was there three days, and without any furniture, when Brian left on a six-week safari. He just handed me a gun and said, 'If anybody tries to break in here, shoot him.'"

All three Nicholson children – Susan, Sandra, and John Philip (named for Ionides) – were born in Nachingwea: Susan is married now, living in Bangkok. "Brian was hunting a man-eating lion down near Mtwara when Susan was born," Melva remembered. "Fourth of August, 1956." But five years later, Nicholson was made Senior Game Warden for the southeast sector of what was now the independent nation of Tanzania, a sector that included all of the Selous, and was posted to the large agricultural center of Morogoro, north of the Reserve, where traders were sponsoring the rampant poaching. As Brian wrote a few years later:

> It is necessary to digress for a while and examine events which were affecting Tanganyika as a whole, for these have a direct bearing on the policies concerning the Selous, the Government's acceptance of our requirements in land, and the availability of the land itself. From the earliest days there never was any official policy directed towards development of Game Reserves. These were simply areas set aside for the maintenance of wildlife, and the idea that they could play an important part in the economy of the country was never seriously considered. This goes a long way to explain the negative attitude adopted by the Colonial Administration towards

wildlife. With a few notable exceptions most officers in Government service looked upon game as a problem – either to cattle or
cultivation – and Game Reserves as useless tracts of land which
could not be used for anything else. Game Rangers were cranks to be
tolerated, and the Game Department was a sort of Cinderella which
was grudgingly allocated inadequate funds and staff to be used
mainly on game control work.

The picture began to change in about 1955 when the impact of
tourism and foreign earnings began to assume increasing impor-
tance to the East African territories. It was apparent from the start
that East Africa's unique wildlife was the main attraction. With the
development of National Parks, and growing local and international
interest in East Africa's wildlife, the Game Department began to
expand and carry more influence. As a result the staff and financial
problems became easier, and the proposals for protecting and
utilizing wildlife were taken seriously, in high levels of Government. By 1958 it was clear that the country was moving towards
independence, and the need to establish a foundation for an
economically self-supporting state undoubtedly had its effects on
promoting tourism, and through this the development of wildlife. In
practice the National Parks got the lion's share of funds and
publicity, but we in the Selous Game Reserve benefited increasingly
from 1958 onwards. At this stage there was still no definite plan for
any form of development within this or any other Game Reserve,
and the idea of strictly controlled hunting was looked upon as
sacrilege. (I had in fact prematurely suggested this as far back as
1951.) The prevailing view at Game Department Headquarters was
that all really good game areas should become National Parks, and
that we were merely the caretakers until that happened. Some of us
did not agree with this policy, since it was perfectly clear that huge
areas of our Reserves were utterly unsuitable for National Park
purposes. It was not until late 1961, when Major B. G. Kinlock
became Chief Game Warden, that alternative uses for the Reserves
was accepted, and the idea of carefully controlled hunting became
policy in the Selous Game Reserve.

Though his responsibilities had now shifted from elephant control
and anti-poaching to conservation and development, Brian continued to
spend as much of his time as possible in the Selous, which had been his
special interest from the start. At this time he was still implementing
Ionides's master plan, and he showed me on a map the boundary
extensions that he had brought about during the 1960s; these included
crucial additions in the north, east and south, most of them designed to
spare outlying wildlife populations from black hunters (who brought in
no revenue), the better that they might be killed by white. By the early

1970s, hunting safaris were not only supporting the whole operation of the Selous (three main Game Department bases, including Kingupira, a staff of 460 people, well-manned game posts, 3500 miles of dry-season track, three ferries, and three bridges) but producing revenues for the government as well: the Reserve was divided into 47 hunting blocks, each with a predetermined, low annual quota for each species, and the 94 hunting safaris that came to the Selous in 1972 paid out 1.5 million shillings in trophy fees alone.

The 1960s and very early 1970s were prosperous years for East African wildlife, when a great amount of money was brought in from Europe and America due to the efforts of able administrators, in particular John S. Owen, Director of Tanzania National Parks. (It was John Owen, an old friend of the Eckhart family, who had introduced me to Maria.) Owen himself and all his wardens suddenly found themselves flying airplanes, as did many of the research biologists attached to the Serengeti Research Institute; the fashion for airborne wardens and biologists spread like bush fire, even to poor relations of the Parks such as the Game Department. In 1969, Brian Nicholson learned to fly when the Game Department acquired some small airplanes, and 24 airstrips were added to the efficient system with which he now administered the Selous. That same year, he was transferred to Game Department headquarters in Dar-es-Salaam as "principal Game Warden and Advisor to the Government on Development Projects" – among others, the planning of which regions of the country would serve best as national parks and which as game reserves. Nicholson promptly arranged for the extraction of what he called "the big thorn in my side" – the Mikumi Game Reserve, a much smaller area north of the Selous which was crossed by a new highway from Dar-es-Salaam to Iringa and therefore wide open to poachers in a relatively populous region. Mikumi was simply not large enough or special enough to deserve all the game scouts needed to patrol it; being so close to the highway and the towns, it would work much better as a national park, an idea very agreeable to John Owen. As Mikumi National Park, it now protects a whole stretch of the Selous's northern boundary (though the two are separated by the tracks of the Great Uhuru Railroad, built by the Chinese, that links Dar-es-Salaam to the Zambia copper belt).

When Africans replaced Owen and his parks wardens with the "Africanization" of European jobs in the early 1970s, the wildlife industry lost much of its international appeal, and the decline of Tanzanian government interest followed swiftly. By 1973, Nicholson told me, "everything was beginning to fall apart." The Game Department funds had been cut off, and there was mismanagement of what was left, together with wholesale incompetence. In March 1973, in desperation, Nicholson and Alan Rodgers submitted a report to the Ministry of Natural Resources and Tourism deploring the withdrawal of financial support for the Selous; the enforced reduction of game wardens, engineers, and mechanics, with the inevitable deterioration of efficiency that followed, had meant – among many other problems – discouragement and increased resignations among the dwindling number of competent people on the staff, loss of all but 10 per cent of the Reserve's vehicles due to mechanical breakdown, and the virtual cessation of patrolling, with the result that the poaching activities brought under control ten years before were on the increase once again. Five months later, in the absence of any meaningful response, and realizing that he "was beating his head against a wall", Brian Nicholson resigned his post and returned to Kenya.

In Nairobi, Nicholson found a job as a charter pilot, and soon acquired all the advanced licenses for commercial airplanes. Brian enjoyed flying, and he was good at it, bringing to it the same intensity and dedication that served the Selous so well for twenty-three years. According to his wife, he did not take a single day off in Nairobi, perhaps because there was nothing there to interest him. At one point Brian was offered a job as manager of a new company that was organizing safaris into the Selous, but with the closing of the Kenya–Tanzania border in 1977 this last chance to work in the Selous evaporated. Except for a few brief flying visits he had not returned here since he left the Game Department six years before, and to judge from his sardonic attitude, he was bracing himself for disappointment in what he was going to find.

III

Maria and I were happy to be back in an old-style green "Manyara" tent, with its rain fly that shaded the metal safari table and canvas wash basin and Maria's grass mat, which was our verandah; the tent gave us a sense of homegoing. The night before, as if to signal our return to the African bush, hyena and lion howled and roared, though not with laughter, and toward midnight hippopotami resounded from their pools deep in the Kingupira Forest. The wistful bird calls of the African night died one by one; soon the ring-necked and the red-eyed doves began to call, and then the tiny emerald-spotted wood dove with its sad, descending coos, and the dawn scream of a fish eagle, the tinny notes of the trumpeter hornbill, the nasal, jeering squawk of hadada ibis—as the canvas filled with light, I lay on my camp cot in the crisp green tent in the greatest happiness.

At daylight came the sound of a fierce whacking from the direction of the mess tent, where one of the Africans, setting table, had stepped upon a black-throated spitting cobra that had taken shelter beneath the ground flap at the door. The pretty, sand-colored creature had shot its head out, and young Kazungu sprang backward and ran off to fetch a panga (a machete) to dispatch it. After transporting the dead cobra nearly a quarter of a mile from camp, Kazungu donned knee boots, green rubber Wellingtons, which he told us he intended to wear for the remainder of the safari.

With Nicholson I went on a "reconnaissance" ("We don't say 'game

(Below) Kingupira camp.

(Above) Leaf-nosed bat. (Right) Impala.

Namakambari pool.

run' here in the Selous; that's tourist talk"), crossing a savanna of coarse grass set about with isolated leadwood trees; the leadwood is a species of combretum named for its durability – a dead tree may stand for many years. Soon we turned west into low terminalia woodland, arriving eventually at Kilunda Pool, a small pan where big green monitor lizards shot off low limbs of baobab into the water. These unexpected pans are usually accompanied by termite hills, and Alan Rodgers has suggested that the red mounds, which are partly subterranean, and also the root systems of the large trees such as tamarind and ebony that grow on them, penetrate the hardpan formed by the clay soils to the abundant ground water beneath. Leaching upward through the fissures, the water attracts animals, which paw in the resulting puddles, making them deeper. They are deepened still further when the hippo and buffalo that come there to roll and wallow carry away a thick coat of mud, and eventually a small pan is created. However, the pan's size is limited; if it gets too big, then erosion of the surrounding land during the rains will silt it up again.

In a dead tree perched a motley selection of large birds: a fish eagle and a black-headed heron, two juvenile saddle-billed storks, and a huge gymnogene or "harrier hawk", with a bare, vulturine lemon-yellow face, that went off in strange weary flight through the bony trees. Beside the pool lay a full-grown lion with full belly and no mane; he got up slowly and moved away, displaying no haste until I jumped out of the Land Rover to get a look at the departing gymnogene. Apparently these maneless lions are rather common in this part of the Selous, and Brian recounted other anomalies of the region. The roan antelope, ostrich, and dik-dik, which are typical species throughout northern Tanzania, are entirely absent from the Selous. The silver-backed jackal has only been recorded three or four times, always near this eastern border, although it is common enough around the settlements outside the boundaries; yet in the Serengeti it is abundant even in areas well away from human settlements. Besides the giraffe found north of the Rufiji, and the puku, a swamp antelope confined to the lower Kilombero, the cheetah is also scarce and localized even on this hardpan plain of the eastern Selous; because of the scarcity of the open country that its usual hunting technique requires, the cheetah has apparently adapted to the prevalence of heavy cover by learning to stalk baboons as leopards do.

The behavior of a widespread species may vary greatly from place to place – for example, crocodiles are much more dangerous in some places than in others (probably according to the availability of fish), and the chimpanzees of Kibale Forest, in Uganda, have not been seen to attack and kill other creatures, as they do at Gombe Stream here in Tanzania, and also in West Africa. Nicholson was always suspicious of scientists at the Serengeti Research Institute who tended to "write up animal behavior on the basis of the Serengeti only, when areas like the Selous were far more typical. All those boffins with their great pulsating brains, selecting facts to fit their precious theories! Years ago, some S.R.I. people wanted to come down to the Selous. I said, fine, on three conditions. First, they must work on projects helpful to the Selous – for example, we needed more information on sable and kudu, and we still do; we don't need to know how high a vulture can fly. Next, that they go out on their research safaris in the rains as well as in fair weather, because animal behavior in the rains is a necessary part of the whole picture. Finally, they were not to go to Dar more than once every three months. I never heard from any of them again." This sort of resentment against the field biologists who come into an area and instruct the old-time wardens about their wildlife is heard rather commonly in East African wildlife circles, but Brian Nicholson is better qualified to speak out this way than most: twenty-five years ago, he published a pioneering paper on wild elephant behavior which led the way for the formal studies that came later and is still highly regarded by students of *Loxodonta africana*.[1] Because of the early end to his education, and because it pleases him to appear rough-hewn, Brian camouflages his ideas behind phrases such as "so to speak", "as *you* might put it", and "if that's the word", but in fact he is very articulate indeed.

On the track ahead, a Gabon nightjar fluttered up like a large chestnut-and-gold moth, only to alight just yards away and vanish in the dry season's brown leaves. A pygmy mongoose, quick and rufous, scampered like a huge shrew across the track, then some striped ground squirrels, long thin tails carried upright and waving slightly to one side, and a troop of yellow baboon, less burly than their olive cousins in the north. Where a new flush of green had arisen from the burnt-out black was a herd of Nyasa wildebeest, larger, paler, and more handsome than the race on the north side of the Rufiji, with roan flanks and haunches where that animal is gray, and a remarkable white blaze across the forehead. Further west, in the *miombo*, a fine big civet cat, started from a clump of tawny grass by the tires of the Land Rover, moved away a little distance before stopping to turn and have a look at us. The civet was black-faced, lustrous in the sere pale grass, averting its head just a little, the better to listen, and going on again when it heard no more than the soft vibration of the motor. The civet is not a cat at all but a large omnivorous weasel, a relative of the mongoose and the honey badger. It eats fruit and carrion as well as small animals and birds, and helps to propagate the fruit trees which it frequents by depositing their seeds in the defecation place that it returns to again and again, sometimes for years; there the seeds thrive, not only because of the powerful fertilizer but because the small rodents that normally eat up the fallen seeds avoid the civet smell and leave the tree nurseries alone.[2]

A group of buffalo went rocking away through the small trees; a lone hyena sat up like a sphinx. On the savanna as well as in the open woods there were impala, which seem to occupy the ecological niches filled further north by the gazelles; in the Selous, the impala is the major prey of the wild dog and the scarce cheetah. In the long grass of the *miombo*, these elegant antelope have the kongoni habit of climbing on to ruined termite mounds in order to see better.

The alluvial hardpan of this eastern boundary region (the hardpan is formed by river clays mixed with old sand washed down out of the eroded soils of the *miombo*) is characterized by terminalia thorn bush; lacking the ground water that is found in most of the Selous, the land depends for its water on the rivers and also the clay-bottomed pans that sometimes hold water throughout the dry season. With Hugo van Lawick, I spent a day at a large pan called Namakambari, or the Catfish Pool, an harmonious place set about with terminalia, albizzia, and combretum, and also small black cassia trees in yellow blossom. Here a hundred-odd hippo were in residence, but as the dry season progressed and the pool shrank they would retire to the rivers and the last deep holes in the Kingupira Forest. The water lettuce at Namakambari had been stomped to a green mat by the hippos (which eat very little aquatic vegetation), and the place was a natural illustration of why most borehole wells created artificially for animals, both wild and domestic, turn out to be such ecological disasters: the grass and ground cover were obliterated by the pressure of all the animals using the pool, and the packed earth, baked hard as brown concrete, extended as far back into the bush as I could walk without losing my grasp of the myriad game trails and becoming lost. In the green rim of trampled lettuce, a small company of waders picked silently along the margin: common and wood sandpiper, ruff, little stint and the three-banded plover with its coral bill – the only one of this far-flung group that makes its nest in Africa. There was just one individual of each species, and probably these birds were not early autumn migrants from Eurasia but a makeshift community of those left behind by the northward migrations of the spring before.

In the early morning, the blue sky with its high cumulus was crossed by big dark birds – griffons and hawk eagles, bateleurs and vultures, as well as the gaunt water birds scared up from Namakambari. A flock of thirty-two open-billed storks soon returned with heavy flapping to settle in a sepulchral arrangement on the bare limbs of a dead tree; the open-bills are so named on account of the odd space between long bent black mandibles through which one may see the sky. As the sun rose, the dark birds crossing the sky returned to earth and the hippos, which had settled somewhat at our approach, lowered themselves deeper still into the thick gray-brown broth of their own making. A gray heron poised in the water was evidence that fish and frogs could still find sufficient oxygen to exist in this copiously fertilized water, and that the water itself could scarcely be deep enough to immerse a standing hippo, far less a swimming one; the enormous animals were resting on their knees.

I sat very still in the thin shade of a tree that grew from an ancient termite hill close to the shore. A *brrt brrt* of wings preceded the arrival of chin-spot flycatchers, and soon other birds came to the bare limbs and dead snags nearby: doves and rollers, a white-headed black chat, the lesser blue-eared starling, sparrow weavers, and a brown-headed parrot that could not make up its mind whether I was something it should investigate or merely flee. On one dead limb over the pool, two hammerkops peeped sadly as they mated; a pygmy kingfisher, turquoise and fire, zipped into a burrow hidden in the mound behind me. Striped skinks emerged beside my book, and the parrot followed me all around the little hill, clambering along on the limbs over my head with electric shrieks of indignation as I stalked a very small deliberate slow bird, modest olive-gray above with pretty gray bars on a white breast, called the barred warbler. Searching for mites, the warbler worked from the base of a small bush up to the top, flew down and started again, always moving upward from the bottom until it had circled the mound to my place once more, where it proceeded to glean the leaves near my right hand.

A herd of impala picked its way around the pool to a point just yards from where I sat; their harsh tearing snorts as they suddenly departed would warn me, I thought, of the approach downwind of any lion. Soon wart hogs came in from the far side, progressing forward on their knees, tails whisking and manes shivering as they snouted and rooted in the baked earth. From the pond, in the thick heat of the growing morning, came a pungent duckpen smell to which the Egyptian geese that swam around at the edges of the hippo herd made only a pitiable contribution. The geese never appeared to feed, seemingly content with the sheer overwhelming presence of their huge and indelicate companions, and they stayed close, retreating only when washed backward, attending minutely to each thrash and heave as the herd barged about in its small space, as if there were much for a goose to learn from hippopotami. Periodically the cacophony of groans and blares, snorts, puffs, and sighs subsided with the submergence of raw, agonized heads, leaving only a mute cluster of shining wet boulders on the still surface of the pool. Then, one by one, the heads protruded, froggish pink eyes and round pink ears, followed by the generous nostrils that can close tight under water.

Sometimes hippos remain beneath for minutes at a time, thinking long thoughts or cooling the cumbersome machinery of their brains, or – in deeper water – enjoying a short stroll over the bottom. But in these close quarters the commotion resumes rapidly, a quake and rumbling from beneath the surface, then a roar and wash as the huge bodies surge, and way is made for two pink-eyed gladiators which draw near slowly, splitting each other's ears with heavy bluster. Sometimes one will turn aside, not to flee but to hoist its hind end out of the water long enough to defecate, the fleshy furious short tail whisking muddy manure into the unoffended face of its assailant. (Since subsurface elimination is much more relaxing – as easy, indeed, as rolling off a log – Hugo has concluded that this strenuous act, which would surely be taken amiss among human beings, is a gesture of submission among hippos.)

Many of the outbursts were not true fights but the threat display of a female hippo, directed at those which approached her calf too closely; this maternal solicitude invariably incited an uproar, though it soon deflated into disgusted snorts and weary sighs, as if to say, "What can be done with such crude people!" Since the animals were all piled up together, the cow appeared to be drawing a fine line, but no doubt she could perceive a threat not discernible to the casual observer. Despite appearances, hippos are sensitive and easily upset; they were not reconciled, even hours later, to the presence of Hugo's car, which they stared at all day with suspicion and pursued with bluster charges toward the water's edge whenever it was shifted or appeared to be departing, in order to speed it on its way.

I noticed, however, that when real fights occurred between two males, the herd did not join in the uproar but fell silent, as if watching carefully for a sign that the hippo hierarchy was about to change. Even the Egyptian geese retired as the gigantic creatures reared up on their hind legs, mouths wide and ivory clacking; their huge heads locked, the titans twisted, crashing back into the water in an attempt to come to grips as a dung-filled wave rolled across the pool, flushing the birds up from the margin and washing the water lettuce with a rich soup of manure. Then a third male came in from the side, in discreet silence, to deliver to one of the straining contestants a terrific bite upon the flank, driving it off. He then turned upon the other and engaged it in a contest of jaws which he soon won. Only when the fight was over did the nervous herd release its tensions with a mighty uproar, as if the opinions of each one had been vindicated, subsiding shortly once again as if nothing had happened. Most of this was ritualized combat, minimizing injuries, as it is among many if not most of the horned and antlered animals, but hippo bulls may be slashed open by the enormous shearing teeth, and often die. At midday one of the vanquished, apparently banished from the pool, came very quietly out of the hot scrub, anxious to get in out of the heat; he stood indecisively on the bank, great head resting humbly on the mud, as if listening for favorable vibrations. If so, he heard none and decided not to risk it, for after a while he turned away and walked back slowly into the bush, revealing a large open gash on his hind quarter.

In the early afternoon I joined Hugo in his blue Land Rover. The car is specially adapted for photography, even to the green net mesh that may be lowered from the roof on the camera side and twined with branches, thereby transforming this no-nonsense machine into a mobile bush. Hugo is a superb observer (it was he who made the famous discovery that the Egyptian vulture is a tool user, having learned to shatter the smooth enigma of the ostrich egg by slinging rocks at it) and he is full of interesting lore about African creatures, from the wasps and mantises around the camp to the huge African megafauna (as scientists would have it) that have survived the Ice Age. Not long ago, he observed a hunting wasp that had injected just enough toxin into a cockroach to permit the dazed creature to be led by its antennae to a hole where the wasp, having laid its eggs, sealed in the prey as food for its forthcoming young, using two small sticks to block the entrance.

Although a very private person who rarely draws attention to himself, Hugo answered dutifully enough when I asked him about his life. With friends in Holland, at sixteen, he had formed a debating club; one day this club took a trip to a national park, where Hugo was handed a pre-set camera in order to photograph the wild moufflon sheep. Being small as well as quiet, he could sneak up better than the others. "I saw this was something I was good at, and decided to learn how to use a camera," Hugo said, with a characteristic look of pleased surprise. He has been sneaking up on animals ever since. After an apprenticeship with a Dutch film company, he went to Africa in 1960 as an assistant to the Belgian animal photographers Armand and Michaela Dennis, expecting to photograph wild animals on safari. Instead, the Dennises put him to work photographing their captive animals outside Nairobi – it was in this period that he trained himself in the study and photography of insects, which at least were wild – and the next year he tried to get started on his own. Meanwhile, he had become friendly with Jonathan and Richard Leakey, who introduced him to their eminent parents, and eventually Louis and Mary Leakey invited this young, broke photographer to come and live with them. He had not been with them two weeks, he recalled, when the *National Geographic* magazine rang up from Washington: they were doing a film on the Leakeys' work at Olduvai Gorge and wanted to know if Dr. Leakey could recommend a cameraman; Dr. Leakey suggested the young man who happened to be standing at his elbow. The *Geographic* liked the film and shortly assigned van Lawick to do another, on Jane Goodall, a young British protégée of the Leakeys who was studying chimpanzees at Gombe Stream. Hugo and Jane were married in 1964, and he worked with her closely on her Gombe project, seeing to all the administrative work and taking part in the behavioral studies. They have a son, now twelve, also called Baron Hugo van Lawick but better known as "Grub". For many years Hugo served as *National Geographic's* man in East Africa, but since 1967 he has operated as a freelance, producing such beautiful books as *Savage Paradise*, about the predators of the Serengeti. For the past twelve years he has lived mostly at Lake Ndutu, on the Serengeti, just outside the boundaries of the national park.

Hugo and I had been introduced ten years earlier in the Serengeti, but had never really met until a few days before. We now discovered that we had both put aside other projects in order to accompany strangers into the Selous for a whole month because for years each of us had been alert for the chance to come here. Only a well-planned safari like this one could expect to penetrate the Selous to the depth that would justify the effort, and such a safari was the ambition of almost everyone we knew who was concerned with wildlife in East Africa. Very few had made it, and we took the opportunity without hesitation when it came.

Maria, who was raised in Tanzania, Karen Ross, completing her studies as an ecologist, and Robin Pope, a wildlife guide in Zambia's Luangwa Valley, were all very eager to come too, and the Nicholson family was happy to return. As for Rick Bonham, he had persevered despite the warnings of people in Nairobi that his caravan would never get across the Tanzanian border and despite the refusal of the insurance companies to underwrite him; like Hugo and me, he refused to miss this chance. Even my publishers in London had informed me that they saw this project not as a commercial venture but as something that "ought to be done". The only one with a cynical view of all this enthusiasm was Brian Nicholson, who could not bear to be thought soft-hearted or sentimental.

Wild dogs visited the pool, first two, then the whole pack. The strange bat-eared creatures circled around behind the car with curiosity, emitting that odd grunt-bark of alarm that contrasts so strangely with their birdlike twitterings of greetings and contentment. These were all good-looking animals, with shining black masks and brindle on the nape and shoulders, glossy black and yellow-silver bodies, irregularly splotched, and alert clean white-tipped tails. All the carnivores we had seen so far in the Selous – the hyena, lion, and wild dog – were big healthy animals with fine coats, entirely lacking the scuffed and tattered character they acquire elsewhere. This may be because the abundance of water and good pasture reduces the need for seasonal "migrations" of their prey and the resultant stress of leaving their own territories.[3]

In the late afternoon the hippo calves began to surface, the small heads appearing right beside their mothers. The calves are born and suckled in the water, and can lie so low, with only their nostrils emerging, that we did not realize how many there were. In the scrub on the south side of the pool I surprised a hare; it ran off from its dreaming place beneath my feet, the low sun shining through its ears. Nearby, a leaf-nosed bat hung in a low thorn bush like a dried fruit. The first appealing bat I ever saw, it had large, alarmed black eyes in its wheat-colored fur and wrinkled wings of golden apricot that were filled with sunset light. In the north a heavy smoke ascended from a great bush fire, bruising the high cumulus with purple yellow, and a pair of hawk eagles were borne down the evening sky on a hard wind; this southeast trade wind is the prevailing wind throughout the year except for a period at the end of the dry season, when it backs around to the northeast in anticipation of the rains. The sky darkened, and yellow cassia blossoms brightened in the dusk.

That morning, Brian had asked me if I wanted to take a rifle, knowing that I had planned to walk around at Namakambari; at the evening camp fire, he casually warned me again. "Even professional hunters sometimes think the hippo is too fat and slow and peaceful to be dangerous – it isn't so." On one safari, he remembered, he had had to shoot three of them, though on only two of these occasions was it the hippo's fault. The third time, on the Luwegu River, which flows down into the great Kilombero to form the Ulanga, his porters had been amusing themselves throwing stones at a hippo that had got cut off in shallow water. After trying unsuccessfully to retreat, the beleaguered beast finally came for its tormentors, who in their panic led it right to Brian. He was sitting on the ground, his rifle beside him, taking tea on his "chop box" – the tin box in which his safari utensils were carried – and when the hippo noticed him and charged, he had to shoot it; it collapsed, he said, with its great head facing him across the box.

On another occasion, Brian told us, he had sat perched on a termite hill "splitting my sides with laughter" as a hippopotamus pursued "the acting chief game warden, Mr. D. Keith Thomas, who was on an official tour of my area" round and round it. Although this story was super-ficially comic, since hippos lend themselves to slapstick, Nicholson knew better than we did that a hippo can bite a man in two, and I found it difficult to believe that even Ionides could split his sides when actually faced with the possibility of such an outcome – the thundering beast and screeching human being about to be bloodily destroyed before one's eyes. Not knowing quite what to make of his story (not to mention his attitude, in case the incident were true), I peered across the firelight searching for some sign of mischief in his face. ("I had a rifle," Brian explained later, "and was in a position to control the situation if it started getting out of hand.")

Andrew Geddes of London, who made an airplane visit to the Selous with Nicholson and Arnold a few years ago, has testified to Nicholson's expertise in what the Warden himself refers to caustically as "eyeball-to-eyeball" encounters with dangerous animals; Geddes has described to me how Brian deflected a charging elephant with a rifle shot that struck it below the eye. "He stepped between us and that elephant; I'll never forget it. He saved our lives." In short, Brian's credentials were beyond dispute, yet I found myself resisting certain details of his accounts. This instinct was borne out by Melva Nicholson, who spoke with loving pride even of those headstrong qualities in her husband that from time to time must have caused her distress, but who was wonderfully frank and outspoken in all matters, whether talking about her own relentless snoring or racking her memory over certain details of his stories. "I thought there was just one lion," she might say, and when her husband would say evenly, "Always two lion, Melva," she would not dispute him, merely nod her head.

In any case, with these hippo stories an uncomfortable silence had fallen on the company, which was relieved only when Hugo told two hippo stories of a different nature. On one occasion, an Egyptian goose was perching on a hippo's back and when her goslings, skittering and peeping at the hippo's side, tried to climb up, the animal, no doubt irritated by the patter of tiny feet, had turned and taken a tremendous bite at them. One gosling vanished into that enormous maw, only to come sailing out again unharmed as the closing jaws expelled a wave of water.

Another day, a male hippo had chased a rival out of a pool and pursued it out of sight over a rise. Soon the earth shook again as the conqueror returned, still traveling at high speed, and hurled himself with a huge triumphal splash – ha-*whum*-pha! – into the water. Perhaps twenty minutes later, the vanquished hippopotamus turned up, moving slowly and discreetly, taking a full minute to ease his bulk into a corner of the pool with scarcely a ripple.

As the days went on, Brian and I got on better than either of us (I suspect) anticipated; all the same, we were still feeling each other out on the sensitive matters of race and politics. One night, over a tot of rum (my tot: he scarcely drinks), I put forth the widely held idea that man's brain capacity had not improved for the last 40,000 years. Until 10,000 years ago, I suggested; all men were hairy hunter-gatherers, perhaps nocturnal, in which case the chances were that they all had blackish skins. Hugo pointed out that chimpanzees had white skin under their hair, which seemed to suggest that white skin was no evidence of evolution of primates. We wrangled a little, rather uselessly, on old questions such as the true definition of "civilization" and the obstacles to "progress" in a tropical environment. Emboldened by drink, I concluded spiritedly that the white man judged Africans by the material standards of his own reckless civilization, by the "progress" that was ruining the human habitat, and threatening the future of the earth . . . ! But Brian, of course, also deplored such "progress", and I lost track of my argument. Abruptly we changed the subject, and peaceably, partly because both of us were talking half-baked nonsense, and partly because we wanted to get along – indeed we would *have* to get along, as it now seemed certain that we would make an extended foot safari into the wild region between the Luwegu and Mbarangandu rivers with no company but the Africans and each other.

Brian was quiet for a time, considering me in a certain way he has, lower lip curled, head cocked a little sideways, eyes lidded and cold.

Subsequently he related an incident at Moshi in 1951, when he was temporarily attached to Game Department headquarters near Arusha, due to the excessive zeal of Nonga Take-Your-Heart. "A herd of elephant got into that banana belt between the plains under Kilimanjaro and the mountain forest, and the Chagga couldn't get them out. These elephant found themselves surrounded by a million Chagga throwing rocks and sticks at them. Then one who was a little smarter than the rest set dogs on them, those frightful little *shenzi* yappers, and those elephants *really* got angry; they went tearing after the damned dogs, and the dogs ran back into the village with the elephants close behind, you see, and they all went round and round among all those new pride-of-the-nation houses with tin roofs, and when the elephants got fed up trying to catch the little dogs, they tore into the Chagga houses, ripped those new tin roofs right off, tore them apart. And after that, they shot off down the mountain and kept going, went all the way across the border into Kenya."

It was a funny story and he told it well, and I had to laugh about the elephants and little dogs; but watching Brian watch me laugh, I wasn't sure that we found the story amusing in quite the same way, and hoped that the differences would not cause trouble on our foot safari.

Author with Brian Nicholson.

IV

After a few days we were to move our base camp south and west about seventy miles to the Madaba region, near Nandanga Mountain, where C. J. P. Ionides is buried; David Paterson would meet us there with the supply plane. Meanwhile, in hope of photographing sable antelope and greater kudu, Hugo, Brian, and I made a "fly camp" safari to the Tundu Hills, perhaps twenty miles from Kingupira, setting off through the wild-dog woods where five of these apocalyptic creatures, half-hidden, watched us pass. At a side track beyond the Kilunda Pool, we turned south to a dry sand river called Chimbulili, then southwest once more toward the open woodland ridge called Nakilala. The half shaft on Brian's Land Rover was broken, depriving it of four-wheel drive, and in the wet shallow grassy valleys that occur so unexpectedly in these parched woods, his machine had to be hauled out twice by Hugo's winch, with the aid of thrust from Bwana Peter, old Saidi, the Chagga boy Renatus, Hugo's mechanic and assistant, and Mwakupalu, the assistant cook, who would tend to the sahibs on this brief safari. For Brian, the broken shaft was a minor frustration compared to the scarcity of animals in one of his old haunts in the Selous. "Always elephant and buffalo in this valley, *always!*" he said. "Usually sable or kudu, too, and often both." But all we saw along the way were two solitary buffalo, two small bands of Lichtenstein's hartebeest, or kongoni, a common duiker, and a band of zebra, very wild, fleeing like striped spirits through the trees. (This is a slightly smaller race of Burchell's zebra of northern Tanzania, with narrower stripes that look black rather than dark brown.)

Here in the Kingupira region, which is relatively open and accessible, we saw most species only once and in small numbers, and most were exceptionally flighty – so flighty, in fact, that Nicholson, who had yet to see an elephant on this trip to the Selous, was already speculating that someone had been shooting at the animals. But, as he said, the situation in the Selous was quite different from that in the famous national parks of northern Tanzania, where wildlife could be readily observed not only because that highland country was more open but because many of the animals were hardened to the hideous sounds and outlandish sights and smells associated with the vehicles that turn up stuffed with human masks and glittering lenses all day long. "The parks are all very well in their place, but they are *parks*," he said. "The Selous is the real Africa. This is what most of Africa really looks like."

In the Selous, the spoor on the tracks and in the stream beds testifies to the abundance and variety of animals, but because of the wildness of the place, one must hunt them out and count each sighting as an event. This suited me entirely. (Rick Bonham agreed. "To me, this is the heart of Africa," he had said. "This is how it used to be. The place is stacked with game, even if you can't see it – signs everywhere, even back in the *miombo*. And what you *do* see, you have all to yourself. In the parks, there's always a minibus parked next to it." This was especially true of the "Southern Circuit" parks near the Kenya–Tanzania border, especially Manyara, Ngorongoro, the Mara Game Reserve, and Amboseli. In recent years the more remote parks, such as Ruaha which Maria visited in mid-August, have been lacking in visitors as well as funds and staff, and buildings, roads, and basic maintenance were breaking down.)

For a photographer the situation was very difficult. Hugo was having trouble getting close to animals, and had to shoot through screens of foliage when he succeeded. Before he came, he had been warned by Alan Rodgers that the Selous animals were wild and hidden, but he had counted on their numbers to give him the opportunities he needed; even the Serengeti, he had heard, could not compare with the Selous in its large mammal populations, if one excepted the wildebeest and the gazelles. But whether this was true or not – and we were no longer confident – things wouldn't be easy. Not that Hugo complained; he was too professional for that. But he missed the conditions of the Serengeti, and I could not blame him. The animals further away from Kingupira might be less nervous, and those in the remote south the most trusting of all, but Hugo could not count on this, and as the days passed I could see that he was worried.

At Nakilala, where we arrived just before nightfall, Mwakupalu made quick tea while Brian paced around, disgusted; he was now convinced that someone had been out "hammering animals" for food, and perhaps for money. This region had recently been burned, and there was no regrowth to attract animals; instead, the black floor of the woods, with its cinder dust and gloom, seemed to emphasize the silence and the emptiness. It was plain to see that there was no systematic burning any more, Brian said, far less the foot patrols that were absolutely necessary if new game scouts were to learn their area. Effective patrols could not be made in vehicles, and anyway, all but the main tracks had been allowed to grow over and deteriorate to such a degree that the majority were now impassable. Without patrols, the bloody poachers could come in here as they pleased, the whole place would be shot to pieces, and meanwhile the bureaucrats, the townspeople who had been assigned by the socialist government to training in wildlife management[1] whether they cared about wildlife or not, had taken over the Game Department from the good people Brian himself had trained. His people were less educated, no doubt, but at least they were interested and committed, with a sense of pride and accomplishment in their work, and in some cases – he pointed at old Saidi – with a whole family tradition behind them. "These lazy people they have now do nothing but sit around down here trying to figure out how they can get themselves sent somewhere else. We were proud of this place, and these people despise it! For them, it is banishment and punishment. And even while they are ruining the place, they are telling their superiors in government how well everything is going – a whole tissue of fabrications!" Brian snorted. So far as Brian was concerned, the Workers' Committees that made it so difficult to fire incompetent people had been fatal to morale in the whole Game Department. "Like so many of these socialist ideas, the theory is all very well, but it just doesn't work." The matter was complicated by the cumbersome bureaucratic structure of the Tanzanian government in which scarcely anyone dared to take responsibility, far less risks, lest he be set upon by his ambitious peers, particularly where the decision involved a white man. As Rick Bonham said, speaking from the hard experience of trying to arrange the logistics of this safari, Tanzanians seemed more "brainwashed" than Kenyans in regard to the perils of cooperating with whites, whom they were apt to obstruct as a matter of course rather than be accused of collaborating with "Europeans". Had it not been for the interest and cooperation of Fred Lwezaula, head of the Game Department, and especially of Costy Mlay, one of President Nyerere's aides who had once been on Nicholson's staff in the Selous, this safari could never have taken place. (When Brian and I called on Mr. Mlay in Dar-es-Salaam in September, I found his intelligent concern for wildlife and the Selous extremely heartening; he understood perhaps better than we did how crucial it was that the Selous be saved, not only for economic reasons and for Tanzania's future but as a counterweight to headlong "progress", to help keep man in balance and harmony with the other creatures on the earth.)

A few months after Brian left the Game Department in 1973, the government issued a ban against all big-game hunting throughout the country, even though the strictly administered hunting safaris in the Selous had supported Tanzania's entire Game Department operation and had earned crucial foreign reserves as well. "They couldn't afford to lose the revenues from those safaris. The Selous can support a hell of a lot of hunting, so long as low quotas are determined and strictly enforced, as they were in my day. With no funds to maintain the place, look what has happened: the airstrips are overgrown, the tracks are going, the game posts mostly abandoned, and poachers – not local meat-hunters any more but organized gangs with precision weapons – are said to be coming in from the roads and settlements to the north." It was even rumored that those gangs might be led by the ferocious Somali, who had looted the Tsavo region of ivory, rhino horn, and leopard pelts, but Brian doubted this. "Bad lot, those Somali, don't let anything get in their way: they detest the Bantu, you see, absolutely detest them. But here they'd have a problem with supplies. They won't sit out there eating rhino meat like the local poachers; they have to have their rice."

In 1978, the government had reinstated hunting in the game reserves, not under the Game Department but under a new national company called Tanzania Wildlife Safaris, which in Brian's opinion was unqualified to do the job.

Over the campfire at supper, we had a mild dispute over the matter of the African's concern for wildlife, which Brian had said was virtually non-existent. But a number of my friends, George Schaller and Iain Douglas-Hamilton among them, have had African assistants who were expert and devoted in this field when they were given responsibility as real participants; and the boy Renatus, who was with us here, became fascinated by wild animals because Hugo took the trouble to instruct him. Hugo felt that Africans were increasingly interested in wildlife, and though he grumped a little for old times' sake, it turned out that the Warden agreed. At Kingupira he had received a letter from one of his former staff expressing sadness and concern about what had befallen the Selous:

> Salaamu! I am well, and if you as well as your wife and children are well I am very happy. I am content to hear that you have come to walk here and I wish to send you best wishes and good luck for your safari . . . But for now The Selous Game Reserve is dead. There are no roads, there is nothing but hard times.
>
> Salaamu to your guests. For their sake I feel sorry that there have been no roads for three years now.

Earlier, Brian had praised the dedication of a former assistant named Damien Madogo, who had been so zealous that when he made his Land Rover patrols he had used two drivers since no one driver could maintain his pace, and of the three old game scouts, Saidi, Goa, and Bakiri, whom he had invited to join us on this safari. Goa, who would arrive in a few days, had been Brian's gunbearer and tracker, a traditional African of the bush who cared about the animals, and that old man – Nicholson pointed at Saidi, who giggled – had been very good, too.

Saidi Mwembesi, who quit the Game Department in 1973 with the title of assistant game warden, was a tall, erect Ngoni Zulu. Like most of the people of southeast Tanzania, who have long since absorbed the Arabs who once ruled this coast, he was a Muslim and wore a white Muslim cap. The other Africans called him Mzee, a term of respect that literally means "Old Man"; when asked what year he joined the Game Department, Saidi looked away into the woods, working it out by speaking as he thought. When finally he turned to present his answer, Brian listened politely, then laughed for the first time that evening. "I like the way he tells the time," he said. "He reckons that he joined the Game Department 'the year that Bwana Niki was ordered to kill the man-eater at Songea, and lost a porter to other man-eaters at Kichwa Cha Pembe'."

Over the fire, Brian began talking about man-eating lions, which used to take about two hundred people in southeast Tanzania every year. (Though this figure seemed high, I have heard a similar estimate from Maria's father, who had to sew up the rare survivors of attack by lions at Mtwara Hospital.) The Songea lion that Saidi had referred to killed an estimated hundred people near that town in 1951, and it sometimes killed more than it could eat; if driven off, it soon returned to take somebody else. On one occasion, having killed a herd boy, it chased three others who were fleeing and killed all three before returning to consume its original victim. The local people had decided that this intemperate beast was a manifestation of a certain man who wished them harm, and they did not wish to incur the wrath of this witch by cooperating in efforts to destroy it, but news of the slaughter eventually got out, and Nicholson was dispatched to Songea. Since there were no roads, he had to walk all the way from Kilwa, a journey of twenty days, only to find that the people would not help him. Eventually, he said, the lion was killed by an African with a spear; the man did not survive the struggle.

On the way to Songea, just south of the Selous boundary, Nicholson's safari had made camp near a place called Kichwa Cha Pembe, which means "Head of Horns". About two a.m. on a humid rainy night, a pair of lions came into camp, circling Brian's tent and investigating the main group of sleeping porters before returning to a thatch shelter behind his tent that was occupied by two men, one of them a porter named Mbambako. Instead of going around to the open side, one lion scratched through the back of the shelter and stood upon Mbambako's companion while taking the head of Mbambako into its mouth. The shrieks of the man under its paws awoke the camp, and people fled hysterically in all directions. "The Africans in this part of the world are terrified of lions, due to all the man-eaters; they used to go screaming up the trees at the slightest disturbance. I jumped up and shone my torch, expecting to find nothing at all, but there they were, consuming Mbambako right in front of my bloody tent."

Surprised by the light, the lions dragged the porter's body off into the bush. The drag marks and blood spoor were easy to follow when Brian took up the trail a few minutes later, attended by an unhappy gunbearer who aimed the flashlight beam over Brian's shoulder. "Came up with them after only fifty yards," he said. "Could have followed them on those tearing sounds alone. Came round a bush and there they were, ten feet away. The lioness had already disembowelled him: she raised her head up and I gave her the first bullet in the chest and she slumped right down. At the shot, the male raised up, he had torn an arm off and he had it in his mouth; I gave him the second bullet, and that was that."

Someone who has seen a man torn to pieces by two lions has a firm grasp on the realities, and since on this fly camp we had brought no tents, I was content that, at Brian's instruction, Saidi should build a second fire on the other side of our low cots and lay the Warden's rifle down beside his head. Nevertheless, Maria and I, and Hugo too, had been mildly surprised by Brian's insistence that people leaving camp on foot be accompanied by an armed game scout, and even more so by the presence of a rifle in every Land Rover that went off on an excursion. (In fairness, I should say at once that three of my friends have had their Land Rovers attacked and seriously damaged by large animals of three different species.) Politely, I ignored the rule about the scout, since the presence of a guard with a loaded rifle would destroy the feeling of intensity and vague suspense that is one of the pleasures of walking about in Africa, and eventually Hugo, too, was able to disengage himself from the armed escort, who took up too much room in a car full of equipment. Just as politely, Brian had ignored our dereliction, though once he spoke to me about Maria: "You must tell her that she must not despise the bush," he said. "The fact that she walks from one place to another without an incident does not mean that incidents won't happen." When I warned Maria as instructed, she laughed. "When *you* stop walking all over the landscape, I might consider it," she said. She was quite aware of the presence of lions as well as of big herbivores, but the reward of entering the African silence was worth what seemed to us a negligible danger – negligible, that is, so long as one remained alert, kept away from high grass and dense thicket, and avoided crowding or cutting off any animal with the capacity to put an end to the unnatural two-legged apparition that had frightened it.

Karen Ross had told me that the previous owner of her car was a young American wildlife researcher in Kenya who was taken by a lion on her usual evening walk between the tourist lodge and the camp where she was staying at Amboseli. According to the ecologist David Western, who

(Right) Buffalo.

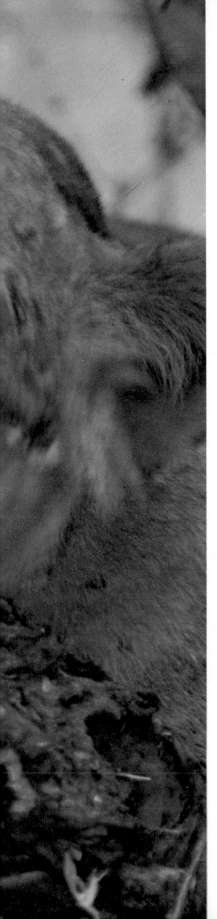

(Below) Nyasa wildebeest.

(Above) Hyena.

(Below and right) Impala.

Greater kudu bull.

later gave me the details of this story, the year was 1973 and the camp had in fact been his own camp – the girl was using it while he was working in Nairobi – and she had been taken, not in darkness, but about one o'clock in the afternoon. Since she lived alone no one reported her absence for two or three weeks, and even then, it was assumed that she had simply gone off to Nairobi; it was not until her car was discovered in the shed, full of rotting food and safari equipment, that a search party was raised, about five months after her death. "One of the rangers found her skull, which was still intact; I was later that week to find her hair, safari clothes, and books. The evidence of scattered books, a few baked tracks, and the damage to the clothes was enough to tell the story. It would seem that she was killed by the lion grasping her neck, but such details are more speculative than real."

From a point just past an isolated bush located near the path to Dr. Western's camp, the trail of dropped books was found leading toward the house; apparently a lion had been hidden by the bush, the girl had walked past it, and the books had been scattered as she fled for the house upon being chased. Since the skull was intact, one must suppose that the poor girl had lived briefly, aware of her own obliteration in the heat of the African day.

Out in the black woods all around, the bush fires were still burning, and the soft African voices came and went on the east wind; during the night I heard someone rise now and again to throw more deadwood on the fires. Until this journey, Renatus was inexperienced with animals except those seen from the safety of Hugo's car, and Mwakupalu, on a recent safari into Amboseli, had been so frightened by park elephants that came boldly into camp that he had cut a hole in the back of his new tent and fled into the bush. Yet it was neither of these two but the old game scout who got up from his blanket by the Africans' fire on the far side of the fallen tree to feed the blaze that warned away the big night animals.

Unknown bird calls, unknown stars of the southern hemisphere – I listened to the keening of a lone mosquito, under the black branches that traced the blue-black heavens overhead. A thousand frogs trilled, and the Southern Cross was lying on its side, far away down the night sky toward the Cape of Good Hope.

Brian had reminded us that in the morning we should shake our shoes out, to make certain that they did not shelter scorpions. Since Brian, perhaps to enhance our adventure, seemed fond of pointing out the perils of safari, I did not take this very seriously. But the next morning, when I sat down on the big log by the fire where Mwakupalu was making porridge, a large and shiny charcoal-colored scorpion clambered out on to the gray wood-ash beside my boot.

◆ ◆ ◆

(Left) Sable antelope.

According to Saidi, who had talked to the game scouts at Kingupira, the track to our next destination, at Madaba, had been allowed to deteriorate so badly that it was now impassable for Bonham's truck; the alternative was a long roundabout route, using a side track off the main route to the south. Under the circumstances it seemed sensible to eliminate the second base camp we had planned there and proceed directly to the far south, making only a side trip to Madaba in order to collect David Paterson. When I suggested this, Hugo promptly agreed; he was anxious to have a long-term base so that he could give any photographic project all the time required. Brian also agreed, after a moment. In any case, we would have to return to the Kingupira track first thing in the morning; it was crucial to cut off Rick Bonham, who was leaving Kingupira for Madaba that day with a lorry load of fuel and food.

We headed cross-country. In a stately wood of tall, well-spaced trees and open grassland from which huge strawberry-colored crickets rose before the tires, a grove of small hookthorn acacias, like old orchard trees, were in white blossom. Toward a big bare leadwood where two coucals were duetting, emptying strange notes from a magic flagon, flew the first new species (a "life bird", as ornithologists say, meaning one never seen before) that I have come across in the Selous – the violet-crested turaco, big, soft blue-gray and forest green, flashing the beautiful crimson flight feathers that are found in most members of this family.

We arrived at the main track toward the south just in time to head off Bonham's lorry. Rick was pleased by the new plan, which would eliminate the trouble of the second base camp; he would take this cargo straight on south to the Mbarangandu River. Anyway, he had bad news about Madaba. A party of American hunters who had leased a Madaba block for a three-week safari had quit in disgust after two days. The source of most of this information was a professional hunter from Zambia, who had come through Kingupira the day before; having spent three weeks in the Madaba region, working with a Tanzanian hunter under the auspices of Tanzanian Wildlife Safaris, the government safari company, he had found only two elephants worth shooting for their ivory. He had also heard that T.W.S. had killed 126 elephants that year in order to subsidize their losing operation, and that none of these elephants had ivory bigger than 50 pounds to a side, which is only average (a very good bull might carry twice that weight or even more). Worse, he had seen what looked like places where helicopters had landed, which seemed to suggest "official" poaching, or that the animals were being killed, as rumored, to help pay for the war in Uganda.

Richard Bonham is a fair-haired wiry young man of twenty-five, slightly strung out by bilharzia, and rather dour in the laconic style of bush types everywhere throughout East Africa, perhaps because he was still exhausted from the stress of getting this safari into Tanzania. "I thought the Selous was the last stronghold, man. And now the poachers are right in there in the very middle of the Reserve."

"Those bastards," Brian Nicholson said. He spoke very quietly, slumped behind the wheel in the empty woods. Then he raised his eyes. "Some of the biggest ivory in the Selous was right there around Madaba, Rick. Discouraging, isn't it? And they must be hammering them in this district, too. I hadn't thought it would be possible to be three days in the Selous and not come across a single elephant, but now I'm beginning to see why."

When Rick was gone, we continued northeast toward Kingupira. But soon Brian turned off northward, arriving eventually at a small sand river, a channel of fine white quartzite sand perhaps twenty yards wide that the old Land Rover, with its broken half shaft, could not cross. In the east wind, big thick purple-red blossoms of kigelia fell silently to the dry ground. With Saidi as gunbearer, the Warden set off to investigate the far bank on foot, and sensing that he wished to think things over, I wandered in the opposite direction, looking at birds. A beautiful red-headed weaver (unlike most weavers, rather solitary) climbed about in a pod mahogany or "lucky bean" with fresh green leaves that danced and glistened in the morning sun, and a parrot preened its vivid green in the fresh parrot-green of an albizzia. The ubiquitous white-crowned sparrow weavers squabbled domestically at woven nests like balls of hay among the yellow blossoms of a cassia, and around the bare clerestory of a great baobab, striped swallow wove fine, unseen strands of sweet, unswallowlike song.

Returning, Nicholson said tersely, "I'm discouraged. It's everything I'm hearing, not only from Rick but from the game scouts. I'm *glad* I left the Selous, you know; if I was trying to run this place in the face of what's been going on, I'd be round the bend." He got into his Land Rover and once again slumped behind the wheel. "You're right, Peter. Let's head straight down to the Mbarangandu. Maybe these bastards haven't been working that far south, and there will still be something left to see." Leaving Saidi and his *bunduki* (rifle) behind to guard Hugo and me from the missing animals, he set off for the Kingupira camp.

On the way home, he told us later, he had stopped off at the Miombo Research Center to see what he could learn from the Game Research Officer as to the whereabouts of the Selous's famous elephants. "The elephants are now declining," the Game Research Officer informed him. "I can see that," agreed Brian. "Why?" "They are dying from some unknown disease. We find them dead." And he showed Brian a whole roomful of tusks that were now the property of the government. Recounting this, Brian rolled his eyes in exasperation; he suspected the worst. But he had already convinced himself that things would be different in the south, and was eager to go as soon as possible. At Kingupira, he had hired eight Ngindo who would help repair the neglected airstrip down on the Mbarangandu River and serve as porters on the foot safari that he and I would make in the far south, beyond the junction of the Luwegu and Mbarangandu. "It won't be like this down *there*, I can tell you," the Warden told me. "The Selous is the finest wildlife habitat in Africa, and the Mbarangandu is the heart of it."

Making our way slowly back toward camp, Hugo and I stopped wherever something caught our eyes. I observed the wood hoopoes and helmet-shrikes and a beautiful violet-tipped courser, while he photographed an impala herd on the black ground of a brown wood, and the white skull of an elephant against the blackened earth.

If Brian was surprised that he had seen no elephants – and Hugo and I had seen them only once – I was astonished that I had not seen a single new species of bird until the violet-crested turaco that morning. But at the Kilunda Pool, on the return to camp, I saw a second, a bird that had eluded me for years; glancing at what I took to be an emerald-spotted wood dove that had just alighted at the pool edge, I discovered instead the very similar blue-spotted wood dove, with its pearl-gray crown and iridescent night-blue on the wing – not an uncommon bird at all, but easily overlooked among its kin.

Meanwhile poor Saidi had been sitting squashed up in the back of the Land Rover, in the heat, and when finally he emerged at camp, he moved quite stiffly. "I used to work hard," he told Maria when she teased him in Swahili. "But now I am old and just sit around letting others do everything – I am good for nothing." And he burst out in delighted laughter, walking away with his rifle, shaking his head.

Whenever Philip or Sandra Nicholson left camp, Saidi went along, unless Robin Pope was free to fill that duty. Saidi's main job on this expedition was to sit about with his rifle wherever Philip decided to go fishing, which he decided to do almost every day. Philip, indeed, was indefatigable, bringing home endless strings of barbels for the delectation of the staff, and occasionally a tiger fish or the naked catfish, which were more pleasing to European palates. Meanwhile, Sandy started a collection of the peculiar natural objects of the *miombo* – the strange hairy seed pods of the crocodile-barked pterocarpus, the winged pods of terminalia and combretum, the beautiful, small, woody pear-shaped pods of the "African pear", and the shiny bronze beans from the long pods of the sterculia. She also struck up a pleasing friendship with young Robin Pope, whose good sense and dependability made him a great asset around camp, although he did not speak Swahili. In his modest way, Robin was very good company, observant and firm in his opinions on natural history – we disputed mildly over the identity of certain species – yet soft-spoken and detached, full of shy humor. Without censure, he described how Tom Arnold and David Paterson, walking through Zambia's Luangwa Valley, occasionally conducted loud political argument from their usual positions at opposite ends of a long line of walkers,

unchastened by the silences and splendors of the Africa they had come so
far to see.

One early morning I walked along the thicket of rapphia palms at the
edge of the Kingupira forest, keeping an eye out for the lions that we
heard each night, and listening to the sunrise plaint of the hadadas and
fish eagles and trumpeter hornbills, the shrieking parrots, the squalling
and explosive *chack!* of four boubous chasing through a bush. High in a
riverain forest tree, four brown-chested barbets sat very still, gathering
heat from the new sun in their crimson breasts, and not far away on a bare
limb, a scarlet-chested sunbird preened itself with staccato energy, as if
dealing with an attack of biting ants. Beside the track, a songbird came to
perch just near the silhouette of what I had assumed was a Gabar
goshawk, which will take such unwary birds wherever it finds them;
looking more carefully, I saw that the hawk was a lizard buzzard.
Interestingly, the lizard buzzard does not take birds, perhaps because it is
too slow and lethargic, and more interestingly still, songbirds have
learned to distinguish this raptor from the others, and will perch beside it
without fear.

At a place where a porcupine had been destroyed, its beautiful black
and white quills lay scattered on the sand. Gathering quills, I heard soft
sounds behind me and turned with a start to see old Saidi, bound on foot
for Ngarambe village, several miles away. I asked him why he was
carrying no *bunduki*; hadn't lions been seen along this track just
yesterday? Knowing that I was teasing him, Saidi gave me a sly look, then
burst out laughing.

V

It was close to noon on 26 August when we broke camp and headed for the south. Not far from camp, in the full heat of the day, five wild dogs were engaged in a wart-hog kill, yanking and tearing at the dying creature in an open grove of small yellowthorn acacia. In the dust and sun and yellow light, among skeletal small trees, the dogs in silhouette spun round and round the pig in a macabre dance in which the victim, although dead, seemed to take part. But within a few minutes the wild pig had been rended, for the dogs work fast, perhaps to avoid sharing their prey with hyena or lion. The strange patched animals loped away into the woods, lugging big dark red gobbets of fresh pig meat in the direction of their den, returning soon to fetch away the last wet scraps, gray now and breaded with dust in the hard, hot wind.

"There's nature in the raw for you!" Melva exclaimed, as unsettled as all of us by the strange scene.

In these closed woods, weighed down by the gray sky, yellow baboons descended stiffly from a tamarind and moved off like old men. The long-faced kongoni looked depressed in the gloom of the burned *miombo*, but a pair of oribi, bright rufous, the first of these small antelope that we had seen, stood alert among the scaly leaves of stunted rain trees, so called because of the froghopper insects that ingest the sap, then spit out droplets of nearly pure water.

The tsetse invaded the Land Rovers and bit hard; occasionally a nerve was hit by a bite of fire. There was much slapping and sweating in the humid heat, and tempers were short. Tom Arnold and Maria in particular had no tolerance for tsetse, and Tom commanded Robin Pope to keep his car windows tightly closed, despite the heat and the poisonous miasma of insecticides.

Apparently tsetse are attracted by the movement of cars and by dark colors; they favor wart hog and are put off by zebra, as was discovered after thousands of zebra had been massacred as "carriers" in early unavailing experiments in tsetse control. Of the twenty-odd species of tsetse fly, most occur in some place or other over more than half the thinly populated reaches of Tanzania, but this *miombo* species is the common or savanna tsetse, *Glossina morsitans,* which has spared a vast area of Africa from human destruction. Due to poor soil or insufficient water or both, most of the *miombo* fails to support a human population large enough to maintain an adequate area of bush in the cleared state that inhibits tsetse, which survive in small pockets and may carry the trypanosome parasites from wild animals to domestic ones, or to human beings; even in well-watered regions such as the Selous, all but a few hunter-gatherers or subsistence farmers are driven out.

Our long day in the tsetse woods was not improved when a Game Department Land Rover honked and passed much too fast, and Brian banged the steering wheel with his palm: "Just look at them! No wonder the machines don't hold up on these rough tracks! That's the so-called anti-poaching unit, I suppose!" He clacked his false teeth, then burst out again. "They'll go all out to Madaba, wrecking the machines, and hobnob with the people who are clobbering elephant up there, have some drink with them, then go on home!"

Toward midday we met Rick's truck on its way back from the south. The track was extremely rough, Rick said, and the truck had mired down; he was very dirty but also very happy and excited. "Beautiful place, man! Ten herds of elephants, big herd of buffalo, lion – the lot!" But before Brian could take heart from this good news, we were overtaken by a Land Rover containing the white Zambian hunter and his black Tanzanian partner, whom he serves officially as an assistant; they were on their way back to Madaba, and stopped to chat.

Brian demanded that the Tanzanian hunter explain the suspicious evidence of helicopters. The hunter, dressed in vivid forest green, acknowledged that helicopters seemed to be visiting Madaba, but looked unhappy and evasive when pressed for details. He denied that Tanzania Wildlife Safaris had killed all those elephants; it was the Game Department, and anyway, only 96 had been taken, not 126. Why, Brian asked, had *any* elephant been slaughtered inside the Selous Game Reserve? Well, actually, a report had come that elephants *outside* the Reserve were doing severe damage to cashew plantations, and the government had ordered one hundred killed. However, none had been located on the cashew plantations, so, you see, the quota had to be met somewhere else! *Why?* Brian repeated coldly. Giving me a baleful stare, he turned away without waiting for an answer and got back into his Land Rover. He had captured a live tsetse fly, and while waiting for the other people to get back into their Land Rovers, he held it between thumb and forefinger and tickled its abdomen. Realizing I was watching, Brian murmured, "For some reason, this makes them swell up – they can't seem to expel the ingested air. Quite interesting, really." And he gave me his bad grin, cheered up momentarily by my disapproval.

The track led on across woods and flat-topped ridges, and by mid-afternoon we came down into the Muhinje Valley, which in other days had been inhabited by a few tribesmen. I rode with Brian and Goa Mwakangaru, a small, quiet man in a straw hat, who had been Nicholson's gunbearer and tracker here in the Selous, and who had joined the safari only yesterday. In the back seat, Goa sat up like a little boy, hands on the seat ahead. He knew this country and was happy to be back here. And it was Goa who called out *"Tembo!"* in his hushed deep voice; his tracker's eye had picked out pale gray shapes deep in the woods. Brian smiled, shaking his head; these were the first elephant he had seen in the Selous, but the event was now so anti-climactic that he could not comment.

In a little while, pointing into some undistinguished woodlands, Goa smiled a little and whispered, *"Maji ya Bwana Niki."*

Nicholson turned his head to look at him. *"My* water?"

Goa nodded. *"Maji ya Bwana Niki."*

And Brian shrugged. "Must have been some place I camped," he said. "I can't remember it. In the old days, we used to name these water holes for the game scouts who found them – gave them incentive to get out and look around. Had a lot of dedicated people then, and more coming up; we'd start them off as porters, and promote the good ones. In those days, the job of game scout was the most sought-after in southern Tanganyika, with good pay and great prestige and a new, smart uniform every six months, with red beret; the job was often handed down from father to son. That tradition's all but gone now. Bakiri says they're not giving out any new uniforms at all. People like Saidi and Goa and Bakiri Mnungu are the last of the old scouts who know this country; this lot today wouldn't dare go off the roads, or they'd be lost."

Goa Mwakangaru was originally from the Tsavo region of Kenya, the son of a subchief of the Taita, an elephant-hunting people from the Taita Hills. As a youth, Goa hunted regularly with bow and arrow, and was credited with three rhino and two elephant. Subsequently, when an anti-poaching campaign led by the Tsavo wardens put most of the elephant hunters in the "Hoteli Kingi Georgi", Goa gave up his vocation to become a tracker for a professional hunter then working in both Kenya and Tanzania. When Goa was in his early twenties, this man recommended him to Brian Nicholson, whom he served until Bwana Niki quit the Game Department and returned to Kenya. "He has marvelous eyes," Nicholson said. "I reckon the sign of a good tracker is the ability to follow a lion over hard ground in the dry season, and Goa can do that. He worked with me once or twice on man-eaters and often on cattle-killing lions, until I moved up to Morogoro; after that, I was mainly concerned with conservation work and the Reserve."

In the valley called Nahatu, we made a fly camp between two big trees, an *Acacia Sieberiana* and a big vitex tree of the verbena family that in season produces a black edible fruit. We drank cool water from the ditch in the bottom of the dry-season karonga and listened to eared owls – the eerie trill of the tiny scops and the hooting of the spotted eagle owl – and slept communally under the kitchen fly, using mosquito netting. Brian and Melva had a tent put up, in which Tom Arnold joined them. Tom made no bones about his nervousness around big animals, and preferred to sleep with a Land Rover nearby. "I wasn't raised to it," Tom said, with a candor that all of us admired.

That morning, as we continued south, five waterbuck in a glade of red-barked afromisia watched us pass, and a shiny hippo rose from a small lake and walked away among low trees in a sedate manner; otherwise, the tsetse woods were still. The track passed through a region known as Nambarapi, or Place of the Sable Antelope, and we kept an eye out for sable but saw none. Nor had we seen any kudu; so far, these two big woodland animals had remained hidden.

At a dim crossroads there were crumbling signs, put up long ago under Brian Nicholson, and not far to the west the famous steam engine – the one intended to grind millet to make bread for the soldiery of Count von Lettow-Vorbeck – sat in the long grass off the track. A brass plate commemorated its construction in 1858, and seeing the size of it, one could scarcely imagine the cruel labor of those Africans who were forced to haul it on its iron wheels through a hostile waste of swamp, thornbush, and karonga, all the way from Kilwa, on the coast, to somewhere west of this place, for it was now facing east; unless it had been hauled backward, it was apparently abandoned on the return journey. Though it has tilted and begun to settle in the sandy soil, this prototypical contrivance will serve for many decades hence as the last monument to "the Battle for the Bundu".

A man who was on Brian's staff when he first came to the Selous had been conscripted as a youth to help haul this steam engine, and the grandfather of Bakiri Mnungu also served in this part of Shamba ya Bibi as a soldier for the Germans. Before leaving his village he declared, "*Naomba Mungu Naingarezi Wasiniue,*" meaning "I hope that God [Mungu] keeps the British from killing me," from which, according to Bakiri, the villagers gave him the nickname Mnungu which has persisted as his family name. Or perhaps Bakiri preferred this spelling and interpretation to the simple "Nungu", which means "porcupine".

At the spring called Mingwea, camp was made in a grove of tall, airy, dark green legumes known to the Nyamwezi of western Tanzania as *muyombo;* this is the *Brachystegia* species from which the *miombo* habitat derives its name. Here the Game Department car that had passed by with such intemperate haste the day before stopped off on its way back to Kingupira, and Brian Nicholson, leaning one elbow on the hood of his Land Rover, his cigarette holder in his mouth and a sardonic look upon his face, studied the eight game scouts whom the car disgorged. Led by a man recognized by the Warden as a former skinner for a professional hunter, the eight smiled warily at the expression on the face of Bwana Niki, and, to my relief, their leader thought better of offering his hand. Still leaning against the car, Brian responded to the greetings of "the General", as he later referred to him ("had bigger and better boots than anyone else, so I assumed he was in command"), by asking what the game scouts imagined they were accomplishing out here. The former skinner, dressed mostly in civilian clothes, admitted cheerfully that he did not really know: they had investigated a report of poachers and, finding none, were on their way back to headquarters. ("Probably expected to find them sitting on the road," Brian commented later.) This man too had heard reports of helicopters and, like the hunter we had met the day before, he appeared uncomfortable with the whole subject. Brian said shortly, "We have to get on about our business," and the eight retreated to their car again.

Remembering Bakiri's information that Game Department uniforms were no longer available, I felt sorry now for the only man among them who was trim, clean, and in full uniform, even to his red beret and lanyard, a good-looking young African who in other days would probably have won the grudging favor of Bwana Niki; his forlorn effort to wear his uniform with pride seemed very gallant.

Before departing, the men told us that they had seen greater kudu near the track, and Hugo and I, accompanied by Goa, headed further west toward the Madaba River. Soon we met another Game Department car, and also a big truck that carried some sort of cargo under a tarpaulin. Considering the fact that the patrol post here had been abandoned, there seemed to be an awful lot of activity in this region, and we wondered if that cargo meant that more elephants had been found dead.

The kudu were gone from the place described. *"Watu wenge,"* Goa said: a lot of people. But kongoni at least were abundant, and there were elephants and, standing just inside the track, the first bushbuck we had seen in the Selous.

At the Madaba airstrip, where David Paterson was to come in that day with the supply plane, we rejoined Brian, Melva, and Sandy, whom we accompanied still further west on a pilgrimage to Ionides's grave on

Nandanga Mountain. "They've let all the tracks go, as you see," Brian warned Hugo, whose Land Rover was a new one. "We'll have to do a bit of *bundu*-bashing." And indeed the track up to Nandanga was entirely overgrown and rutted, with dangerous potholes, and so dim in places that Goa had to get out and hunt around in order to follow it at all. No effort had been made to clear even the smallest trees; instead, the few drivers who had passed in recent years had detoured around through the woods and on to the track again.

On a high open ridge that overlooked a wide expanse of the Selous, Brian stopped to point out landmarks to the south and west, and that note of homecoming excitement came back into his voice again. "That's Mberera Mountain, where the Kilombero River comes down to the Luwegu at Shuguli Falls to form the Ulanga River; from here, the nearest point of the Ulanga would be just over there, due west, behind that ridge. Further north, the Ulanga joins the Ruaha and becomes the Rufiji, which is of course the largest river in East Africa, and it's just north of there that the Rufiji turns east toward Stiegler's Gorge." Melva Nicholson had got out of the Land Rover and was staring around her at the wide lowland spread beneath, with its swamps and water courses and stretches of savanna, and now she cried out, "*Nothing!* Used to be *full* of animals down there, and now there's nothing!" Her husband had been saying the same thing in so many places for so many days that he was sick of it; he scarcely nodded. My binoculars could not turn up a single animal, and after a little while Brian said mildly, "Haven't been here for ten years now, perhaps more. I must say, I hadn't expected such a change."

In 1860, the German explorer von der Decken, who gave his name to the common hornbill of this region, complained about the lack of game in this southern country – very likely, a direct consequence of the ivory trade and the great slaving caravans that were still passing between Kilwa and Lake Nyasa. The ivory hunter Jim Sutherland took some elephants along the Luwegu River in 1912, but traditionally they remained scarce in southeastern Tanzania until the advent of the British, whom the elephants were popularly believed to have accompanied from parts unknown. Yet in the 1930s more than 2500 animals were shot each year in the southeast on elephant control, and as late as 1961, 800 were shot in Kilwa and Utete Districts alone as an encouragement to their companions to avail themselves of the peace and quiet of the Selous Game Reserve, where sensible elephants might retire when they got fed up with human administration, instead of just trumpeting off across the land.

Even so, adventuring elephants still wander out of the Selous, and in trying to account for elephant scarcities, it is well to recall that elephants come and go according to whims and vagaries of their own, disappearing for years and even decades from likely areas, only to colonize the place again when least expected. In 1880, contesting the commonly held idea that the supply of ivory was inexhaustible, the British explorer Joseph

Thomson took note of the sad state of affairs in what is now Uganda: "In my sojourn of fourteen months during which I passed over an immense area of the Great Lakes region, I never once saw a single elephant. Twenty years ago they roamed over those countries unmolested and now they have been almost utterly exterminated." Eighty years later Murchison Falls, now Kabalega, in the north part of "the Great Lakes region", was the site of the first elephant-cropping program in East Africa. In the 1890s, the ivory hunter Arthur Neumann encountered no elephants at all while crossing the entire extent of the Tsavo country; in 1970–71 more than 6000 elephants died at Tsavo on account of drought and degeneration of their habitat caused by over-crowding. In 1913 the first safari into the Serengeti found no elephants at all; in 1968, with Myles Turner and George Schaller, I saw more than 500 in a single herd.

In an air survey made in 1976 by Alan Rodgers and Iain Douglas-Hamilton, the estimate arrived at for the Selous was 100,000 elephants, about a third of the Tanzania population (which, with Zaire, claims half the elephants in Africa). Douglas-Hamilton says that this figure may have been too low since many animals must have been missed due to the rough nature of the terrain: "If you want to be conservative," he told me recently, "just say 'over one hundred thousand'." The survey noted a remarkable lack of the habitat damage so pronounced in most of the parks, probably because the elephant were broken up into small groups, mostly five or less, widely distributed throughout this vast, trackless and well-watered reserve – which I like to think accounts for the fact that on the ground one might encounter rather few of them, even along the river margins. But to judge from what we had seen so far, of course, it is too high. Brian Nicholson says that in the 1960s, he and Alan Rees, at that time warden of the western sector of the Selous, arrived separately at the same figure of 30,000, using a ground-survey technique worked out between them. Nicholson however has complete faith in the competence of Alan Rodgers and feels that the survey must have been more accurate than his own figure, since he and Rees could only extrapolate from a rough sample taken in a relatively small area.

If Brian's uneasiness about their scarcity is well-founded, then they may have departed of their own accord for elephant kingdoms in other parts of this vast country. Thus we may hope that in the far south the hidden thousands will appear. At supper the night before someone had spoken of a recent novel in which the last herd of *Loxodonta*, fleeing the insatiable guns of blood-crazed *Homo*, hid themselves in a huge swamp by walking in there backward, in order to persuade their pursuers that they had departed from that place.

On the far side of the Madaba River, the track passed a stack of junk and rusting fuel drums; collapsed and splayed out in the undergrowth lay the tin roof of what once had been a hut. "In my time," said Brian, "there was a permanent patrol post here – four men, rotated every three months. Old Bakiri Mnungu used to be in charge here. He's been at Kingupira for the last three years, and he told me he still doesn't know that country; in three years, he said, he hasn't yet been sent out on patrol, not even once." The Warden gazed about him bleakly. "They're running out of people who still know the bush; Bakiri and Goa are among the last. All these new people do, when they do anything, is run up and down the roads wasting petrol and beating up the machines, like those ones that you saw this morning. Not their fault, of course; there's nobody in authority any more who is interested enough to set them an example."

The Nicholsons drove on, and Hugo and I followed, feeling very subdued. But as the track continued to deteriorate, pushing ever deeper into the bush, kongoni, wart hog, wildebeest and zebra began to appear in encouraging numbers, with a few waterbuck, buffalo, and elephant, another bushbuck, a bush duiker, and two bull eland, one of them a lordly specimen of august hump and swinging dewlap and thick spiralled horns. The game was especially abundant on the further side of a deep sand river called Kipilipili, a steep-banked barrier to vehicles which was only crossed with the help of Hugo's winch; we hauled his machine out of the sand bed, using a tree, then winched out the other vehicle, which had got buried to the axles trying to bull its way up the steep bank. There were more bones of dead elephant than seemed natural for one locality, and in one place two skeletons lay together, suggesting that the beasts had died at the same time. Both of the big skulls had worn-down teeth, an evidence of age which might have meant large tusks. "That's what's happening, all right – bastards! And that so-called anti-poaching unit, running out here at top speed all the way from Kingupira, wrecking the machines and turning right around and going back –!" Realizing he was repeating himself, the Warden left the rest unfinished, shaking his head.

Here and there flew the lovely racquet-tailed rollers that seem to replace the lilac-breasted species in *miombo* woodland, and in a big pterocarpus, bare except for its odd, hairy seed pods, sat a dark chanting goshawk, clear gray with long red legs. The track had deteriorated entirely, and at the next sand river, a tributary of the Kipilipili, we ascended the white sand of the bed until the stream was blocked by a fallen tree.

From this place, we headed off on foot, and almost immediately Goa sprang backward, smiling shyly in apology for his alarm; he had heard a noise from the high grass on the bank. "Probably a puff adder," Brian said. "Come round this way, Melva." We detoured around the grass clump and started off again. Brian walked just behind Goa, and both of them carried rifles. Sandy stayed close behind her father, followed by Melva, who set off bravely but was out of condition; like her husband and most other East Africans, she is a heavy smoker. Try as she would, she soon slowed

Brian Nicholson at Ionides's grave, Nandanga Mountain.

down, very red in the face, in the thick high grass and humid heat and the steepening hill. "I do wish my legs weren't so short," she sighed. Melva suggested that Hugo and I carry on without her, but we didn't wish to leave her behind: elephant and buffalo dung was fresh and plentiful, and the grass in places was well over our heads. Melva tried again, and again slowed down. The Warden was now far ahead, not looking back, though Sandy did so, a bit worried; Sandy herself was more hardy than she looked and was going strong. "He'll *never* stop," Melva said despairingly. "Not for *anybody*." In this complaint there was a note of pride, as when she says, "He never takes a day off, *never*." Desperate, however, she called him and, when he did not seem to hear, called again – "Brian!" Sandy turned around, obviously concerned, but she did not chide her father, and I found this odd; his impertinent pretty Sandra was the only one from whom the Warden would take teasing, and she needled him constantly and to his great delight until she got him laughing, as she said, "like a hyena". Seeing that Melva was done for, I decided to yell at him myself, and this time he turned, and they awaited us on the open hillside under the big, dark-trunked *muyombo* trees.

It was already mid-afternoon, and we had to keep moving. Melva was left in the care of Goa as the rest of us ascended to a foothill ridge under a steep red cliff face of Nandanga Mountain, which lies very near the geographical center of this vast reserve. Nandanga is only 3000 feet high, but it has power, like any isolated monolith arising from low plains; it is an outcropping of basement rock from the great African shield, some of the most ancient rock on earth. On the summit of the ridge stands a handsome stone inscribed with the name of Constantine John Philip Ionides, 1901-1968, "erected with great respect by members of the East African Professional Hunters' Association".

"Old Iodine!" Brian said. "*There* was a hunter! He had joined the British Army in the twenties just to go out to India to hunt, and when he heard that the hunting was better here, he transferred to the King's African Rifles – that must have been about 1926. When he quit the Army, he became an elephant poacher in the Congo, and finally he signed up with the Game Department, which in those days was mostly concerned with protecting the Africans and their shambas from elephant and lion.

"Iodine lived to hunt, and he was a superb hunter, with great patience and knowledge of the animal. Took him four months of stalking in the Aberdares before he came out with his first bongo! That animal bays up very quickly to dogs, but Iodine refused to use dogs or salt licks or any of the other tricks that are used today. Getting a bongo later became a sort of ambush, all set up for the client by someone who had a particular bongo all staked out; these so-called hunters went out and got their bongo in a single day."

From the grave site there was a prospect of the Mkungu Mountains west of the Ulanga River, as well as the great southern distances of the

Selous. Looking at the headstone, Nicholson grunted, "Iodine always said he wanted to be thrown out to the hyenas, but that if I *had* to bury him, it should be here."

As in his attitudes toward human beings vis-à-vis snakes, there was a suggestion of posing in Ionides's attitude toward his own remains: "I strongly object to being a nuisance after I am dead," he had told one interviewer. "I've been a carnivore all my life, and I'd much rather benefit a few local vultures and jackals." And to another he had said, "One of the most stupid of all premises is that life is in some peculiar way sacred, and that every body must have some sort of ceremony performed over it. Nature is an adequate arbiter in these matters. My hyenas would dispose of me satisfactorily." His fond mother – perhaps the only person, and certainly the only woman, he was ever known to put himself out for – confided to one of his chroniclers that her son "was ruthless by nature, absolutely ruthless." It seems mildly surprising that he confessed a wish to rest his bones here, since in neither of the two published accounts of his life with which I am familiar is there any important reference to the Selous; it seemed as if, when he left the Selous in 1954, he had lost all interest in his life's greatest accomplishment. Not that Ionides thought in terms of making a contribution; indeed, he liked to boast of his own selfishness: "I'm completely selfish. Whatever I've done has always been for my own ends and my own enjoyment, because that's the only way to live, in my view – and who else's view is there any sense in living in, for God's sake?"

We gazed about us for a little while at the vast wilderness all around: no sign of man, no marks or sounds, for hundreds of square miles. Even the ring-necked doves had fallen silent. There is a silence in the imminence of animals and also in the echo of their noise, but the dread silence is the one that rises from a wilderness from which all the wild animals have gone. In the dead still afternoon of the old continent, it seemed to me that the silence of Nandanga still had imminence, a listening, a waiting in the air.

The day was late, and we descended quickly to Goa and Melva, then found our way down through the woods. It was important to cross the Kipilipili before dark. In a green meadow bathed in the humid light of the sinking sun, a family of bush pig was setting out on the evening forage. The big boar was gray and his mane silver, but the sow and all the spry young shoats were rufous red, with clean white manes. The boar sensed something that did not belong here, and while his family moved out of sight, poking and snuffling as they went, he stood motionless halfway up a bank, squinting toward the dim shadows at the wood's edge.

At the sand river, walking downstream to the Land Rovers, we followed the old spoor of a lion, and crossed leopard pug marks and the three-toed prints of rhino, which looked like strange traces left by the last dinosaur on earth. Before leaving, Goa set fire to the high grass on the

(Right) *Tree frog on a baobab.*

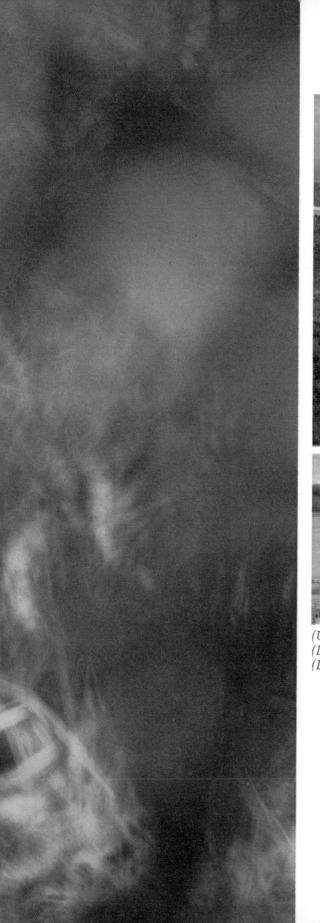

(Upper) Mbarangandu River.
(Lower) Richard Bonham and crew.
(Left) Female impala.

banks, and the fire leapt away toward Nandanga Mountain. Hugo and I stared at Brian, mildly astonished; in some part of his mind he was still Warden of the Selous.

At sunset, in single file, seven wild striped horses galloped away over the fire-blackened ground against copper leaves, and a bush duiker bounded straight into the ball of sun that touched the horizon of the ridge. Then it was heavy dusk, and quickly dark.

Once on the main track we still had thirty miles to go. The headlights wavered along on the rough road, and a thick-knee struck the windshield and went fluttering off ghost-like into the blackness. A skunk-sized creature with white bushy tail, fleeing the light, ran ahead of us for a hundred yards before darting off the path – this was the mustelid called the zorilla, a lesser relative of the civet. Not far from camp, a spotted eagle owl sat on the track, eating a snake without troubling to kill it; undaunted by the huge night eyes of the car lights, it bit at the small, writhing creature in its talons, then lifted its blood-glistened beak, its yellow eyes, to return our stare. At last the owl rose softly from the track, carrying the shining snake into the trees.

Wild dogs killing a wart hog.

(Left) Wild dog pup.

VI

In the early morning of the next day's journey, while the tents were dismantled and packed into the machines, Mzee Nzui, the head cook, tidied the kitchen. He buried tins, papers, and other *takataka*, even the plucked feathers of a guinea fowl, under the ashes, then dismantled his dish rack of saplings and fronds; as Maria remarked, Mzee Nzui was much more conscientious about litter than the whites, who were sentimental about the landscapes of the Old Africa while littering it with their cigarette packages.

Since no one was ready I set off on foot, leaving instructions for them to pick me up along the road. Already the sun was shrouded over with dank cloud, and miasmal humidity had settled into the dark woods. Tsetse abounded, though they did not bite. My footsteps in the soft sand of the track made no sound, and in the windless heat and utter stillness the ominous chinking of a tinker bird, the signal of the black-headed oriole, the see-saw creaking of the coqui francolin deepened the silence. None of these birds showed themselves, and the leaves hung limp as bats in the gray, damp air. Letting my eyes fall to the ground, I saw big round lion spoor, very fresh, implanted on top of tire marks made the evening before. Retracing my steps a little I found the place where the lion had left the track, and realized that it might be watching me at this very moment, that I might have passed it. If so, I was cut off from the camp. Gazing about, I listened attentively, though for what I did not know: it seemed to me that the woods looked rather gloomy. For want of a better plan, I continued on my way, and eventually I heard a Land Rover's quiet hum. Brian picked me up, and we went on past the steam engine, to where the Liwale–Madaba track met the faint track to the south.

Crossing the high grass savanna woodland, we saw scarcely any game at all; not until the track came down into the shallow valley of the Matandu River, a small stream that winds through sour grasslands of black cotton clay, did animals start to appear – a young bull elephant, a yellow reedbuck loping along the wood's edge, then two more by the river, then zebra and hippos and the common antelopes. But the numbers were small, and the sable and greater kudu remained hidden, and Brian became increasingly disgusted. Here, he felt, the scarcity of animals could not be blamed on poachers. The Matandu marks the eastern border of the Selous Game Reserve, but the country beyond the boundaries had been emptied of people years ago in the resettlement schemes set in motion by Ionides. No, it was all this unburned grass that had driven the animals away. These tall tussock grasses produced new tissue only during the rainy season; once they had matured and flowered, they turned dry and stalky and lost all nutritive value. If the Selous were to support large numbers of game, the grass had to be burned off every year. Plainly, the Game Department was not doing its duty. Whenever we paused, Saidi or Goa wandered off the track, and soon there would come a harsh snap and crackling as the land took fire, and the tall pale andropogon grass leapt into fierce color, and black smoke rose behind us in the northern sky.

Knowing that Hugo and I resisted all this burning, Brian spoke impatiently about the views of a certain ecologist who believed that the annual burning as practised in Nicholson's day would eventually transform that part of the Selous into a desert, since fire turns the soil surface to hardpan by burning off the thin layer of humus until only acacia and other thorny inhabitants of arid bush are able to penetrate to the moisture below. "As a scientist, he has to come up with a theory, and he'll find facts to fit that theory," Brian said. In his opinion, quick "cold" fires set not long after the rains, when the conflagration was limited by lack of fuel, encouraged fresh growth without harm to the soil; what did real damage was setting "hot" fires too late in the dry season, when a great mass of accumulated dry grass, stalky brush, and deadwood caused deep burning.

In most places, in any case, there was no humus to burn. The pale ferralitic soils of this central plateau were derived from weathered sands of quartz and granite, which derived in turn from the ancient pre-Cambrian rocks of the African shield; leached out by millions of years of sun and tropic rains, they had become acid and infertile, and the glades of rank savanna grass in shallow valleys and in the water-logged valley bottoms called *mbugas* produced poor forage; where drainage was poor, the *mbugas* turned to clay known as "black cotton", which cracked wide open in the dry season and swelled into heavy morass during the rains.

These poor soils without nutrients, together with the tsetse fly, discouraged permanent human habitation, which is why this vast wilderness had survived into historic times unchanged by man.

Nevertheless, as Brian said, man created the *miombo*, which depended upon his fires to survive; without fire, *miombo* rapidly reverted to a dense thicket. Alan Rodgers agreed that the *miombo* was a recent habitat type, no older than the last pluvial period, perhaps 12,000 years ago, in the lost centuries when the first bush fires were set – accidentally, perhaps – by the early hunters. The use of fire as a hunting tool is very ancient, and its constant use in this uncompromising climate, where the dry season extends almost unbroken for more than half of every year, led eventually to the ascension of fire-resistant hardwoods – in this region, airy, graceful, thornless trees, well-spaced and of fair size, belonging mostly to such characteristic genera as Brachystegia and Julbernardia. This open savanna woodland is the most characteristic aspect of the Selous.

Throughout most of its considerable extent elsewhere in Tanzania, and in Mozambique, Zambia, Zaire, Angola, and Zimbabwe, the *miombo* is flat, dry, and monotonous, a seemingly limitless scrub waste without landmarks or water or other relief for the oppressed eye. Because of the low water table across most of this African plateau, there is little grass regeneration after fires, and the aspect of most of the *miombo*, with its blackened ground and burnt small leafless trees, its humid heat and drought and tsetse, under heavy skies, is immensely oppressive. In typical *miombo*, birds and animals are few, not only in species but in numbers, although the sable antelope and Lichtenstein's hartebeest have developed as endemic *miombo* species, and the roan antelope and bush duiker are more common in this habitat than any other. All these antelope are well adapted to long grass and to browsing; the buffalo, elephant, eland, impala, wart hog, wildebeest, and zebra, all of which are fond of short-grass grazing, are only found in the *miombo* in small numbers. In the Selous, however, there is considerable variation in land form as well as habitat and also an abundance of good water; even toward the end of the seven-month drought there is no point more than six miles from permanent water. As a result – or so I had read, so I had heard – the fauna of the Selous is probably more diverse and more abundant than in any comparable area of Africa.

In the deep woods not far south of the crossroads sat a Bedford truck all but split in two by the pterocarpus tree it had encountered at high speed on this rough track (so rough, in fact, that more than once, we had got out and poked about on foot in order to find it). In silence, Brian Nicholson in his floppy hat, cigarette holder clenched between his teeth, walked all the way round the defunct truck, which had "Selous Game Reserve" painted on the door. He got back into his Land Rover before speaking. "The last time I saw this truck was in Dar, in 1973," he said at last. "Brand new that day, and it was nearly brand new when they did this to it, in July of that same year, just before I got fed up and packed it in. Before I left, I gave explicit orders to have this lorry recovered; haven't got around to it yet, apparently. Look at that motor, and the tires; still perfectly good." He turned away. "This is where all the aid money goes; you give them millions, and this is what they do with it. You should have seen my road grader after just three months; if you'd hired an expert, you couldn't do that much damage!" I thought about the machine-shop-cum-garage I had seen at Kingupira, all the abandoned vehicles, the rusting parts left out in the rain and sun, and no sign whatsoever of activity. It wasn't for lack of good African mechanics – Renatus was very good indeed, and so was Charles Mdedo of Rick Bonham's staff, and Brian boasted of the mechanics he had had in the Selous. No, it was the absence of responsible direction, the absence of *purpose*, Brian felt, that had caused the good mechanics to depart. Seeing his face, I could guess what he was feeling: the splayed truck, with its title on the door, represented twenty years of wasted effort.

Not far beyond the dead truck, on a hillside, there were kudu – a bull and four cows and two young, bounding along a barrier of silver deadwood at the edge of the wood. Safe in the dappled sunlight of the trees, the big animals turned to look as one of the calves began to suckle; then these shades of silver brown seemed to evaporate into the shifting light.

At Tanda ya Nyama (Animal Pool) open-bills and saddlebills, herons, sandpipers, and plovers had gathered, and a male jacana with the powder blue brow of the breeding season led his hen across the water lettuce on long spidery toes. A small hippo clan of a dozen fatty heads observed man carefully as he ate his lunch, and serenaded the Land Rovers with a vast rumpus as we departed once again, heading south into red rock hills with broad prospects of the Mahoko Mountains.

In the afternoon I rode with Karen Ross, and drove her Land Rover. The feel of bush driving came back to me quickly, the ceaseless gear shifting, the easing over humps and trenches, potholes and stone rivers, the bundu-bashing through the trees where a track is blocked, the bucking climbs up steep eroded banks; to minimize the jolts and the crude grinding protest of the car, a certain deftness is required, especially if the passengers, with no steering wheel to cling to, are to be spared.

One of the passengers was Mzee Nzui, our magisterial cook, whose gat teeth in a pale, clean-featured Kamba face made him look less African than Mongol. Nzui's quiet economy of motion, his serenity, even the simple efficiency of his kitchen – the sapling rack covered with green fronds that is used for drying pots and dishes, the wooden hook for moving kettles on and off the fire, the long neat ash bed of the fire itself – reminded me of another Kamba cook named Kimunginye, who in 1970 was cook on a safari which I accompanied to the Northern Frontier

District and Lake Turkana. When Karen asked him if he had ever known Kimunginye, Mzee Nzui's face broke wide with surprise and pleasure; not only was Kimunginye his friend, he was the brother of Mzee Nzui's wife!

Karen was firm and easy with the Africans; she enjoyed their company and sat with them and laughed with them and talked in the same soft flowing way that they do. Therefore they liked her and respected her, especially Mzee Nzui, who could be peremptory and stern in the way he ran his kitchen, and once reprimanded her for trying to hurry supper. Nzui said, "You do me an injustice, Memsahib. I must go at my own pace." But most of the time he gazed upon her with forthright affection, and his feelings were shared by the others on the staff.

Even Charles Mdedo, a sophisticated city man who worked ordinarily for Cooper Motors in Mombasa and might have been expected to resent any criticism from a white woman, gave way to her that afternoon without the smallest rudeness or resistance. Though a competent mechanic, Charles was not used to makeshift repairs out in the bush, and tended to panic without all the equipment and spare parts of a well-equipped garage; that day he was riding in the Land Rover driven by Robin Pope when it broke down, and he promptly announced that the clutch was shot before having a good look at it. By the time we came up, Hugo's mechanic, the gentle Chagga mission boy Renatus, had scrambled eagerly beneath the Land Rover, and now he emerged with a big smile and the good news that the problem was nothing more than a bleeding nut that had vibrated loose, causing the transmission fluid to run out. When Charles denounced Renatus's diagnosis harshly and sarcastically, Karen stepped in very quickly, saying, "Oh, come on, Charles, *pole pole*" – take it easy – and Charles instantly subsided.

Renatus, shamed by the rebuke, stood by the car with his head sunk on his chest. But Renatus turned out to be right, and a few days afterward, when Hugo's Land Rover caught fire – a spark ignited the gasoline with which Renatus was cleaning the engine – Charles was the first to run to Renatus's aid, despite the risk of explosion and serious injury, and was also seen to comfort Renatus later. Putting aside his status and prestige as a mechanic, Charles also helped out cheerfully in the kitchen, even waited at table, and this willing attitude, like his acquiescence to Karen Ross, had nothing to do with subservience or eagerness to please.

Karen Ross is a handsome graceful girl whose father, born Finn Miller Rosbjerg, ran away from home in Denmark and eventually became a naturalized British citizen, changing his name to "Ross" before his marriage. During the Mau Mau days, he read a newspaper article about the murder of a farm manager near Nanyuki and, on impulse, wrote to the farm's owner offering himself as a replacement, provided that his journey out to Africa was paid for. Under the circumstances, he had no competition for the post and his offer was accepted. Upon arrival he sent his young English wife a photograph of the pretty farmhouse, and she agreed to join him, only to find that where they were to live was not the main house at all but the dead man's cottage, a scruffy two-room place with bloodstains all over the floor. Eventually, Ross got his own farm in the Aberdares, and when this was appropriated by the Kikuyu at the time of Independence, he bought a second at Nakuru. "When I was a child," Karen said in her soft voice, "we had lots of animals there, buffalo and elephant and leopard, and I loved the animals and the bush. For a while I thought I might become a vet, but later I turned back to the wild animals. I'm a bush girl, though my family weren't safari people, but I don't think I really appreciated Africa until I was sent away to school in England. I went wild back there, they thought I was some sort of jungle creature. Used to eat up the headmistress's roses, and disappear up trees." In September Karen was to return to Europe to complete her studies at the University of Edinburgh; she hoped to do her thesis on the ecology of Shimba Hills, south of Mombasa, a small game reserve where the only sable left in Kenya still may be found.

More kudu were seen, in fleeting glimpses; buffalo and elephant sign was abundant on the track, and also the big heart-shaped prints of sable. From the southern distance, under odd round-topped hills, rose the white shine of broad sand rivers, then the glint of water as the gray and ghostly stubble of the *miombo* scrub high on the ridges turned into fresh red, green, and copper woods. Where the track descended the last slope of the rivers, the woods on both sides were littered with fresh buffalo dung. At the bottom of the hill, by the ruin of the abandoned Mkangira game post, Brian and Melva Nicholson sat in their Land Rover, gazing out over the confluence of the Mbarangandu and Luwegu rivers, and Karen and I also stopped to contemplate one of the loveliest views that we had seen in Africa.

The Mbarangandu rounds its final bend under steep bluffs on this north side, where the ridges level out on to the plain; on the south side, across the river, lie grass banks and brakes, then open woodland that soon begins to climb into the hills that separate the Mbarangandu from the Luwegu. Seen from here, the Luwegu scarcely curves, seeming to fall straight out of the southwest as if it had descended without bends the entire one hundred and fifty miles from its headwaters east of Songa. (In fact, however, as we saw one day on an air survey made in the supply plane, the Luwegu flows first in a northerly direction, then east, coming swiftly down out of spectacular gorges in the Irawalla region of the Mbarika Mountains and carving broad bends almost all the way to its great junction with the Mbarangandu, where it turns north again toward Shuguli Falls.)

On broad deltas of white sand numerous water birds flew back and forth about their business, and bands of waterbuck lay on the margins like the tame and stately park deer of old paintings. Beyond the tall borassus palms on the far side of the shining waters and white sands rise the blue hills: from this place, the Reserve extends more than one hundred miles to the southwest beyond those Mbarika Mountains, but there were no tracks beyond this point of rivers.

"This is the first time I have ever come here and not seen elephants from this spot," Brian said mildly, complaining out of habit now, with a kind of perverse satisfaction. But he had seen a large buffalo herd as it crossed the track, and a cursory scan with my binoculars turned up two separate groups of elephant.

At our Mkangira camp, which was a half mile further down on the Luwegu River, Rick Bonham reported a good number of elephant and buffalo, and although Brian did his utmost to conceal it, it was plain that he was happy and excited. "Tea ready, Melva?" He knew this camp site from the past, and while the tea was being prepared, he showed Sandy where she might bathe in safety behind silvered river logs that would protect her from the crocodiles.

Filthy with dust and humid sweat, caked with insecticides from the long days in the tsetse woods, we sank down happily in the warm shallows. Sandpipers, skimmers, plovers, and kingfishers moved up and down the brown Luwegu, hippos disported in two different herds within sight of camp, and yellow crocodile and sunset-red impala shared the alluvial edge across the river.

"Where's my tea? Fulfil your wifely duties, Melva!"

"'Tea ready, Melva?'" Melva said. "That's all he ever says when he comes home! 'Tea ready?'"

"Just trying to be sociable, Melva."

And when Melva had served him "a nice cup of tea", the Warden stretched out in his camp chair with the greatest satisfaction, a rare grin breaking out upon his face: "You're a long way from anywhere *now*, I can tell you! The Selous is the finest wildlife habitat in Africa, and the Mbarangandu is the heart of it!"

Before sunset, the diurnal birds were still; only the fish-tailed drongo was still flying. The water dikkop sang its sad, descending song of twilight, and nightjars left their camouflage of bark and leaves to settle on the warm sand of the tracks. At dusk, the tiny scops owl began its trilling, and toward midnight, a fishing owl at the water's edge, not far upriver, gave a strange, low, lugubrious grunt that was heard occasionally throughout the night, though after two days of human presence this shy bird must have gone away, for it was not heard again.

VII

At sunrise, a pair of big male hippos squared off in the river just in front of camp, as airy skimmers lifted along over their fevered brows. One contestant was dark brown, the other flesh-colored. I favored Old Brown over Big Pink (who looked new and raw, and a bit vulgar) although I sensed that Brown was going to lose; as in mankind, it is ordinarily the weaker individual that makes most of the noise, in Brown's case a cacophony of fearful groans and blarts and roars and grumbling, interspersed with deep watery gurgles. On the far bank, a yellow crocodile lay nerveless as the dead, coldly oblivious of all this hippo nonsense, as a pair of small falcons – African hobbies – watched for big flying insects from their high perch in a dead tree, and a few impala, hind legs kicking nervously, stepped discreetly to the water's edge, and wandered away again. The hippos hurled waves of water at each other, bluffed and skirmished, huge mouths wide, then sank from sight, perhaps to make the other nervous. But each time Old Brown went down, he surfaced again a little further off, or faced ever so slightly the wrong way, as if something upriver had captured his attention; he appeared to be giving signals that, for all of his continuing uproar, he had lost interest in leadership and might abdicate gracefully if young Pink, stout fellow, would not hold out for total victory, would not insist on driving his old boss out of the herd. Now both hippos sank again, the river flowed on; somewhere below, I thought, Old Brown was considering a surfacing maneuver that would bring his hind end into play, thus confronting his opponent with a delicate ethical decision as well as a big faceful of manure. The titanic argument went on for several days, until the night when something huge, presumably Old Brown, came running through the camp with thunderous blows of big round feet. The next day, when Big Pink led the herd a short distance upriver, Old Brown was left alone, still and silent as a rock in the sinking river.

In early September, Brian and I intended to head up the Luwegu, crossing over eventually to the Mbarangandu. The rivers were still very high, for the dry season came late, but if the Mbarangandu subsided enough in the next fortnight to permit Land Rovers to travel on the exposed sands, Rick Bonham would try to come upriver, bringing supplies, which would permit us to explore still further south; if there was no sign of him, we should return downriver.

To get a rough idea how far a vehicle might get, we took a Land Rover upriver that afternoon, accompanied by Simon, one of the young Ngindo recruited at Kingupira; the other Ngindo were working hard to restore the Mkangira airstrip under the direction of Bakiri Mnungu. Simon was smart, able, and willing, and Rick described to Brian and me how a few nights before when the truck had got mired on the way down here, the rest of the crew had been satisfied with some makeshift preparations to extract it, but Simon had said, "No. If we're going to do a job, let's finish it properly!" Brian grunted, in mild disbelief. "When you find a chap with that attitude, you don't let go of him," Rick said. "I'm going to try to take him back to Kenya."

Though the broad Mbarangandu, several hundred yards across, was nowhere more than three feet deep, that depth was unusual for this time of year; Brian had heard from the game scouts that there had been heavy rains in late spring. Crossing the river to the broad flat on the far side, the Land Rover got stuck in mid-stream, and having extracted it by jacking it up and pushing hard, we continued another mile or so over the sand to the point where the next crossing would be made. There we abandoned the whole idea of Land Rover travel on the sands of the Mbarangandu: we would carry enough for a foot safari of ten days.

Because it was a lovely afternoon, we went on for a mile or more, wading upriver. Far upstream at the next bend, an elephant moved peacefully, walking on water. "Not bad ivory, that!" Brian remarked, then adjusted his enthusiasm quickly. "Nothing special, of course, not much more than fifty, I should say." Large groups of waterbuck nuzzled the green haze appearing on the damp flats of the river, and two lionesses which had been lying on the far end of the flats trotted away over the sand to the high grass of the riverbank as Simon, excited, cried out, "*Simba! Simba!*" At the mouth of a small river of white sand that came in from the north, a heavy lion spoor included many neat prints made by cubs; there were also the fresh pug marks of a leopard. Not far up the sand river, feeding placidly in rank green swale, was an elephant cow with a juvenile and a young calf, and Brian, seeing calm elephants close at hand for the first time since he had arrived in the Selous, sat down on a high bank and watched them in contented silence. He is obviously very fond of elephants (I am, too; indeed, I think that anyone not fond of elephants cannot be of sound mind) and observing them, he permitted himself a fleeting smile, pursuing some elephant reverie or other, perhaps of the sort described in his own writings:

> As already mentioned, elephants graze a great deal during the early rains . . . When feeding on this new grass they appear to get into a very contented frame of mind, and by moving slowly, even where the only cover comprises four-inch grass, it is possible to approach to within fifteen yards without much risk of being seen. The reason they do not see one is, apart from the above explanation, because when grazing they either close their eyes completely or else look down on to the ground at their feet, when all one can see of the eyes on close examination are the top eyelids and the beautifully long eyelashes.[1]

After some minutes, the cow's trunk stiffened as she got our wind; the trunk rose slowly in an awkward question mark. Then she hurried her young into the cane at the edge of the riverain thickets, and in moments the vast animals were gone.

Barefoot in cool shallows on clear sand, we wandered upriver. Overhead, a white-crowned plover – a brown bird when alighted and a white one when it is flying – mobbed a kite that had circled out over the sands, but the kite was ignored by the flock of green sandpipers, twelve or more, that was feeding along the edges of the bars; this was a migratory flight returning from Eurasia, the birds having completed the breeding season and raised their young. Sneaking up close to see them better, I was startled by the explosion of a hippo from a small side channel near the bank, not twenty yards away. At this range, one is very much aware that excepting the elephant, the hippo is the largest land animal on earth, and since I was between it and the main channel, I was damned glad that this one knew there was no deep water left in the Mbarangandu; it made for the thickets, where its huge shining hindquarters soon disappeared.

Water dikkops flew from beneath a stump and crossed the river, and a young crocodile, dirty yellow with black stripes on the heavy tail, thrashed off the bank and, seized by panic, skittered rapidly along over the surface before subsiding. Though we did not speak of it, all three of us were feeling that here on the Mbarangandu we had arrived at last in the Selous. In contentment, we strolled slowly back along the south bank of the river, pausing to investigate the twin scrapes that the rhino makes by rooting and kicking while scattering its dung. Africans say that God sewed skins on all His animals with a big needle and, becoming tired, asked the rhino, which was last, to do it himself; the rhino did not make a good job of it, which is why it has so many loose folds and wrinkles, and furthermore, it swallowed the needle with the job half-finished, which is why, still searching for the needle, it keeps on scattering its dung.

The hippo does the same, though its reasons differ. When it asked God if it might take up residence in the water, God refused, declaring that an animal so enormous would eat up all His fish and water plants and spoil His rivers. The hippo promised it would feed on land, and to this day waggles its dung about on bushes just to prove to God that it is going properly about its business.

Since it seemed quite likely that lions with small cubs were in residence where we had seen them on the river, Hugo decided he would spend the day there. Maria went with him, and Renatus, and old Saidi with his faithful rifle. Saidi was apparently disappointed that so many of the *wa-Zungu* (Europeans) on this safari were knowledgeable about animals and had no real need of his expertise; he was delighted that this young African, Renatus, was so eager to learn. Although Renatus could get a well-paid mechanic's job almost anywhere, he was happy to remain with Hugo, who not only told him about animals but was very gentle, like himself, and did not raise his voice even when their new blue Land Rover burst into flames. "I am lucky to know him," Renatus said, and Hugo felt the same. While the *wa-Zungu* busied themselves with notebooks and cameras, the old African and the young one watched animals together, and Maria reported that when the day became hot, and the animals were dozing in the shade, Renatus went straight off to sleep with his head on the old Ngoni's bony shoulder.

That day the lions never appeared, but elephants came and went across the water, and Renatus was delighted to see a pair of Egyptian geese drive a fish eagle off their young, not just once but twice, the second time after the eagle had already seized a gosling: the disgruntled bird finally flew away and sat down beside the water, composing itself for a little while before deciding to give itself a bath.

Having pointed out this lion place to Hugo, I walked back to camp, perhaps an hour away, crossing the plain on the north bank of the river. In the early light, the fiery eyes of the lesser blue-eared starlings shone as bright as the yellow cassia blossoms in which they fed, searching out beans from old brown pods that had not fallen. On this windless morning of heavy, humid sun, I noted a shifting in the tall grass just off my route, perhaps thirty yards ahead; I would have been grateful for the cool feel of a rifle. I went a little wide and kept on going.

At camp I described the episode to Rick Bonham, just back from the bush with Robin Pope and David Paterson; they had shot a buffalo to feed the camp. Like the rest of us, Rick had been puzzled by Brian's insistence on armed escorts and on guns in the Land Rovers, but in his opinion Brian was probably right. "Men like that who have had encounters with a lot of animals have learned not to take chances; they know how very fast things can go wrong. He doesn't do it to dramatize things or anything. My father was a game warden, too, and he felt the same; you never caught him out without his rifle."

Rick's father, who came from a good English family, had run away from school and fetched up eventually in Kenya, where he found a job managing the coffee plantation of an old-time elephant hunter named Bill Judd. "It was Bill Judd who taught my father to hunt," Rick remembers, and Jack Bonham kept right on hunting after Judd himself was destroyed by an elephant. He also did some gold-mining, became an ivory poacher, worked in animal cropping in the years when it was thought that wildebeest and zebra spittle in the grass was causing abortions among Maasai cattle, and did some lion-hunting for the Kenya Game Department, which later hired him to crop elephants along the Tana River. Eventually he joined the Game Department as a warden, assigned to the Tana River and the coast, and helped to make a game reserve out of the Shimba Hills, south of Mombasa. About 1950, Jack Bonham had been crippled when a cow elephant knocked him flat with a blow of her trunk, tried to tusk and kneel on him, stamped on his leg, then picked him up and threw him into a thornbush. It took his porters five days to carry him in to Mombasa Hospital. Eventually his injured leg was amputated, and in 1971 he died of a thrombosis apparently related to his old injury.

That morning Rick brought me up to date on the wildlife situation in Kenya, which has undergone dramatic changes. Though Rick himself used to hunt a lot, he thought the hunting ban that was announced in May 1977 was absolutely necessary. "They had to do it, because things had been wildly out of control for a long time. Everybody had elephant licenses – they were taking them out for their grandmothers – and the licenses were expensive, but the price of ivory made it worth while. And for every legal elephant taken, four or five others might be shot by the same gun for the black market, and all this in addition to those being taken by the poaching gangs; they reckoned that only about 2 per cent of the elephants killed were taken legally, on license. Hunting elephant was banned in 1974, but the illegal hunting kept on going, and tons of ivory were still being exported, mostly to Hong Kong.

"Of course it was said that putting a stop to hunting didn't help much, that a complete ban on the trade in skins and ivory was the only way to control poaching. And it was true that months after the hunting ban those curio shops were as full of stuff as ever. But actually it did help a lot, because all the licensed firearms had to be turned in and that accounted for most of the amateur poaching. In the beginning, things got worse, because the professional hunters had known their territory and had kept a lot of the bush country clear of poachers; now the gangs moved in and operated as they pleased, and nobody in government seemed to care much. It wasn't like the old days when the Tsavo wardens were chasing after a few Kamba and Liangulu armed with bows and arrows. A lot of these new poachers were common bandits, especially the Somali around Tsavo; they weren't just killing elephants and taking ivory, they were looting villages and shooting up manyattas and shooting people, too, if they resisted. It wasn't until those Somalis killed Ken Clark, down at Galana, that the government realized that it had to act."

On the day Ken Clark was killed, I was co-leading an ornithological safari out in the Maasai Mara, which in that year of good rains was green and beautiful and overrun with wildebeest that had wandered north out of the Serengeti; I believe it was that day, 3 August 1977, that our lucky clients heard leopard and saw lion, hyena, cheetah, and a wild-dog den with fifteen romping pups, all in one day. But down on the Galana Game Ranch, where Clark, a former professional hunter, served as manager, a gang of poachers was resisting arrest in a day-long skirmish that ended in Clark's death. As Rick said, a number of local people had been killed before that, but nobody ever paid much attention to the deaths of a few peasants out in the bush, least of all their fellow Africans. With the death of a white man, however, and the rush of publicity that followed, the extent of the poaching industry and its connections to people in government, including President Kenyatta's family and the highest officials in the Game Department, could no longer be concealed, and belated action was taken against the poachers, in the fear that the tourist trade so crucial to Kenya's foreign exchange might soon be threatened. In a show of force, several hundred armed men were dispatched to the Galano–Tsavo country, and some armed Somali were summarily shot in the hope that a few of them, at least, were the guilty parties. Just four months later, in December 1977, President Kenyatta decreed a ban on any further trade in animal parts, including ivory, giving Kenya's proliferating curio shops – an estimated two hundred in Nairobi and two hundred others elsewhere in the country – just three months to get rid of their stock.

With Kenyatta's death in 1978 came a symbolic end to the uneasiness in black–white relationships that had followed Independence: his successor, Daniel Arap Moi, regarded the survival of the national parks and the tourist industry as more important than the color of the warden. It was too late for David Sheldrick, who had died in 1977 not long after his removal from Tsavo East, but the present warden, Joseph Kioko, is working very effectively with the former warden of the Aberdares, Bill Woodley, who has now been assigned to Tsavo West, while Ted Goss, once warden of Tsavo West, assists a Kenyan Somali, Mohamed Adan, in the operation of a mobile airborne anti-poaching unit. With the new laws and anti-poaching measures, together with a few years of good rain, the game throughout Kenya has recovered so fast that there is already talk of re-opening the licensed hunting for certain plains game. But almost all of Kenya's parks have seriously deteriorated in recent years – beautiful

Marsabit has been so encroached upon by tribesmen, and poached so heavily, that few tourists bother to go there any more – and could deteriorate further again with the next drought. In the thirty years of its existence, Tsavo East has been transformed from dense woodland to what looks like desert by excess elephant populations and drought, compounded by mistaken management efforts such as digging boreholes, which concentrated the animals and caused them to destroy the vegetation; both elephants and rhinos were already suffering a drastic decline before the poaching gangs came in to finish them off. When I visited Tsavo in 1969, both Tsavo parks had sheltered an estimated 7–8000 rhinoceros; ten years later there were thought to be no more than 180, with a figure of 1500 for all of Kenya. But now the ground cover has been restored, and some young baobab have started to appear. "People exaggerate the destruction of Tsavo," Douglas-Hamilton says, "because they don't understand the extreme effects that shifts of climate have upon that thornbush country."

Amboseli had also been turned into near-desert, though the removal of the Maasai cattle, together with the years of rain, restored the vegetation, and all of the animals substantially recovered – except the rhino, of which there were just ten left. Both Samburu and Meru were seriously beset by poachers, and for several years such northern species as the Grevy zebra and reticulated giraffe were in danger of extinction. Even the Mara, though still prosperous, was invaded by rhino and leopard poachers, and was also threatened because much of it was suitable for growing wheat; north of the Mara, the Mau Mountains were being deforested and the animals killed off by farmers.

"Kenya's population is still shooting up, fastest in Africa," Rick Bonham told me, "and they'll all want land, and the white man's lands have already been taken." He shrugged his shoulders. "I see Africa deteriorating every year, this kind of life, at least." He gazed out over the Luwegu toward the blue hills beyond. "It's such a delicate balance, you see, especially when the governments are so unstable. It's the human population, if nothing else; there's not enough room, so the animals have got to go. That's why my family insisted that I get my commercial pilot's license; I'll have that piece of paper just in case the wildlife situation falls apart and there is no more hope for this safari business."

It was as a pilot for a Nairobi charter outfit, on a three-month job in the Sudan, that Rick got to know another pilot named Brian Nicholson, who spoke to him often and longingly of the Selous; in some ways, this ex-warden reminded him of his father, and he was grateful to Brian for having recommended his new safari company to Tom Arnold. For Rick, as for Hugo and for me, the Selous was a symbolic place, "the last stronghold", as Rick put it, of the wildlife of East Africa. "I'm just happy to be out in the real bush. In Kenya, I don't miss the hunting much except for bird shooting, and that surprises me, because for a while there, hunting was my life. On the other hand, I kind of enjoy this chance to go out and shoot the odd animal for the pot."

I asked Rick what he thought of the Kenya Wildlife Clubs, started as a school program in the early 1970s by a young American named Sandra Price (and recently extended to Uganda and Tanzania). "Doing a fantastic job," said Rick. "That's the only hope. Those kids marched on Nairobi during the campaign to close down the damned curio shops, and they made a hell of a difference. More and more these days you hear Africans talking about wildlife with real feeling. It really means something to them. It isn't just *nyama* (meat) any more." Since the arrival of the safari in the Selous, Rick had gained weight and color, and at this moment he spoke with rare animation. Then he shrugged, gazing out over the swift Luwegu. "I'm afraid I'm a bit pessimistic," he said. "The Selous is really the last hope."

The man in charge of Rick Bonham's equipment was a former elephant hunter and tracker named Kirubai, one of a number of Kenyans who lost a former way of life with the ending of hunting safaris in his own country. Kirubai still served as a tracker when Rick went out to stalk buffalo to feed the camp, and one morning I accompanied them on a short expedition to the far side of the Mbarangandu, where they hoped to find a few guinea fowl for the sahibs' table. But the guinea fowl, so common and noisy when not needed, were off in the thickets keeping their own counsel, and we took time to observe instead the beautiful Boehm's bee-eater, the crowned hornbill, and a beautiful small falcon called Dickinson's kestrel, none of which I had seen before in the Selous.

Eventually Rick shot a dove, and Kirubai, who had collected and discarded a whole series of small sticks, shaved one to a dull point, then pressed this point into the flat shaved surface of another. Using a pinch of the sandy soil for friction, he spun the upright stick between his hands with such concentration that sweat leapt out on his creased brow. When a shallow hole was made, he cut a nick in the bottom stick so that the sawdust could spin out; the spinning continued until smoke appeared, with the spark that would light a tinder of dry grass tucked in below. The dove was skewered on another long stick sharpened at both ends, with the heavier end stuck into the ground at just the angle to suspend the bird a few inches above the embers.

Contentedly, observing hippos in a small river lagoon just beneath us, we cut the dove into three pieces and ate it in the sun, as Kirubai exclaimed over and over at certain subtleties of hippopotamus behavior that heretofore he had not noticed. Rick Bonham thought that Kirubai's experience and ease with the wild animals might have saved his life a few nights before when the truck got mired on its way down here to Mkangira. Kirubai was lying down, digging out the wheel, when he

(Right) Masked weaver.

[96]

(Above) Hugo van Lawick at the Luwegu.

(Left) Egyptian geese with gosling.

White heron.

The Mbarangandu River.

Another view of the Mbarangandu River.

realized that what he had thought was cool night mud was a small snake that had crawled into his shirt. Instead of panicking, he stood up slowly and pulled his shirt out, letting the snake fall before jumping back. "A few minutes later," Rick said, "he was right back there under the truck. Maybe it's because I hate those bloody snakes, but I was amazed."

Further on, Kirubai pointed out what his people call *tuwanda*, a small multi-trunked tree with patchy bark that "bow and arrow people" (the Sanye, Liangulu, Kamba, and Taita peoples of the thornbush country of southeastern Kenya) used for making their bows, wrapping a carefully cut piece in fiber before curing it in the heat of a low fire. Like Goa, who is a Taita, Kirubai had once been a hunter with bow and arrow, but now this tradition was almost finished, and his people were very few and scattered, except for Kirubai's own group, near Lake Balissa on the Tana River. Kirubai himself was married to a woman of Goa's tribe, and was actually Sanye: like Goa, he sometimes identified himself as "Liangulu" because that tribe was romantically identified as "the elephant hunters" by the white men who were offering the jobs.

Kirubai liked the Selous because of all the water here, enough for animals and shambas both, he said; that was not true of his own country, where a few years ago the Somali were finishing off the elephants. In fact, they *boasted* they would finish them, Kirubai said, and furthermore, they were stealing food and clothes and livestock from his people, and killing any who resisted. Now the anti-poaching units were strong again, they were using airplanes, and anyone found in the bush without good reason was thrown in jail, so he thought there was hope now for the elephants of Tsavo. But the only elephants left in Kenya would be those found in the parks, and even in the parks they would not be safe unless they stayed where there were many people. The old days were gone, Kirubai said, and would not come back.

Like all young hunters (Kirubai does not know his age, but Rick figures he must be about thirty-five) Kirubai was trained early to track goats over hard ground – for goats often got lost – and to distinguish between a nanny and a billy. At the age of about eight, his father taught him to stalk and kill dik-dik, then wart hog, using a small bow and poison arrows; whenever he did not learn a lesson, he was thrashed with his father's *kiboko*, or knobbed stick.

When a boy succeeds in killing his first wart hog, he takes the fat home, where his mother shaves his head and rubs in the oil. *Natoka katika kundi ya wanawake*, Kirubai said – You leave the herd of women. The boy is now an apprentice hunter, and must kill a buffalo. Having done so, he returns to the fire in the hunters' camp singing a song – and spontaneously Kirubai sang this song, which included an uncanny simulation of the buffalo's puff of warning and alarm. The buffalo meat is carried back to the main camp for a feast and further ceremonies, which include a second shaving of the head. The time has come when he must

(Left) Baboons.

kill a rhino, lion, and elephant, in any order that they happen to present themselves, after which he may call himself a hunter.

When the boy kills his first elephant, the ears are removed and holes cut in each one so that during the ensuing celebration the ears may be worn upon his arms, together with a bracelet made of the animal's tail bones. Otherwise he is naked. On this occasion, he sings another song:

> I have killed an elephant
> I have cut its ears and tail
> I am a man.

In the still woodland Kirubai sang this song, and I had a pang of recognition: it seemed to me that I had heard a song much like it on another continent, sung by an American Indian. I wondered if long, long ago when, as Indians say, the peoples were all one, this sad brief chant had not been known to all hunters on earth.

When the young hunter takes a wife, he must kill an elephant with tusks more than three *mkono* in length, a *mkono* being an arm's length from elbow to bunched fist. The right tusk goes to the bride's father in first payment, the left tusk is kept by the hunter; sometimes a bride price of five tusks is demanded, as in the case of Kirubai's own mother. But those days are passing, Kirubai said; his mother was probably the last person in his village who still knew how to make the arrow poison. The hunting that the white men now call "poaching" is all but finished, too, although there had been a small recurrence since 1977, when so many trackers and gunbearers lost their jobs that a few felt compelled to revert to their old profession. Kirubai himself had been an elephant hunter toward the end of the old days, when the anti-poaching campaigns at Tsavo put so many of his people into jail, and he was content that he had had that great experience; indeed, he still considered himself a hunter. Among the bow-and-arrow tribes, boys are no longer taught the hunting ways and are forced to attend school instead; no doubt they can tell people their age, as Kirubai cannot. "They are not men," said Kirubai, turning his back on us as well as them and going on into the forest.

In stalking elephants with bow and poison arrow, the hunter had to sneak up very close so that the arrow would penetrate the heavy hide; he shot and ran. When Kirubai was about ten, his father was attacked by a buffalo and his back was injured, and within the year, because he could not run fast any more, he was destroyed by a wounded elephant near the Galana River. In relating this story, Kirubai, upset, acted out the entire episode, pointing his knife at trees and bushes in hushed tones, jumping backward, spinning around as if to hear, and staring fearfully past Rick and me as if we were not there.

With a deep fierce frown, Kirubai told also how his uncle had been killed by a hippopotamus that had torn loose from some wire snares.

His uncle had tracked the hippo to a small pool near the Galana, where people coming in search of him next day pieced the tale together. Apparently he had shot at the hippo, but the arrow had been deflected by a twig; the hippo had charged out and run him down, biting him through the right shoulder and right hip, after which it had returned into its pool. It also attacked the searchers, who fled unhurt; they returned next day and killed it. Coming upon the hunter's bow, they located the tracks of his crawling and followed them to the bush where his body lay.

VIII

On his first day in the Mbarangandu region, Hugo got photographs of a herd of sable antelope; he was red in the face with sun and a bout of malaria but shyly pleased. "That takes care of the sable, I should think," he told me, referring to our book, "but I may have trouble getting kudu." Including the fleeting glimpse he had had back at Madaba, he had now seen greater kudu on six different occasions, but this wonderful creature was as well-camouflaged as it was wary. The animals were not uncommon here, to judge from all these sightings and from tracks, yet Hugo had just one exposure that might work, if it wasn't ruined, as he thought, by too much heat haze. Because he was patient, he was bound to catch up with them as they grew used to his car; at the same time he knew that a month is not much time to get good pictures of shy animals in such thick cover.

In search of kudu and more sable, we returned to the dry *miombo* ridges from where we had had our first glimpse of the broad white sands of the Mbarangandu. The day was heavy, overcast and humid. (Brian Nicholson thought that this morning shroud was the consequence of summer bush fires; it burned off in the late morning every day, with hot blue skies all afternoon and stars at evening.) Perhaps the animals also felt oppressed, for few were moving in the early hours, though big colorful birds were very common – green pigeons and brown-headed parrots, hoopoes and wood hoopoes, racquet-tailed rollers, a violet-crested turaco, a pair of African harrier eagles on a high nest. Just after midday, along the edge of the wood, we encountered a fine sable antelope, grazing the new grass on a burn not fifty yards away. Because the wind was in our favor, the big black animal did not scent us, though he reared his harlequin head as if to listen. Soon he was joined by three more bulls that stepped one by one across the track; one was still young, chestnut-colored and slighter in horn and body than the others. The animals grazed peacefully on the bright green tufts under the fire-coppered leaves of a blackened rain tree. Then the air shifted, the armed heads jerked to taut attention, the white-blazed faces turned to stare even as the shining bodies gathered and sprang away across the hillside, whirling up the dust from the black ground. And as they streamed between the trees, it was easy to see why the sable (and the roan) were given the generic name *Hippotragus*, "horse antelope", not only because of their size and strength but because, unlike such peculiar relatives as the wildebeest and the kongoni, they move with the elegance of horses, lifting their hooves high, heads high too, chins toward their chests, as if to accentuate the grand sweep of the curved horns. There is none of that odd bouncing gait, called "pronking", that is seen in lesser antelopes, even the gazelles. That morning an impala had pronked away into the woods in inappropriate response to the sudden appearance of our blue Land Rover, and kongoni tend to pronk as a matter of course. The long-faced kongoni, with its striped ear linings, blond hind quarters, and rumpled horns, seems suited to this foolish gait; at times it appears to lose all forward motion and bounce straight up and down.

The wildebeest is also a born pronker. On the ridge flats, in low scrub, more than one hundred of these flighty animals now came together, jostling and crowding one another as they pushed forward for a better look at man. The lead bull had imposing horns, which glinted in the sun like horns of buffalo, but such horns are ill-suited to a long sad face with odd ginger eyebrows. The wildebeest has a goat's beard and a lion's mane and a slanty back like a hyena; the head is too big and the tail too long for this rickety thing, and Africans say that the wildebeest is a collection of the parts that were left over after God had finished up all other creatures.

On two high points of the ridge, where petrified wood and small red stones litter the ground, we found a number of flaked stones left behind by the tool makers of early man. A few such sites have been reported in the northern Selous, but so far as we know these sites near the Mbara-ngandu are the first to be discovered in this southern region. I found one of several chopping tools of dark red stone, with flakes chipped off all around the perimeter; it was nearly round, and of a distinct type that Hugo, who had done field work with the Leakey family, had not seen before. "Actually, that is a beauty," said Hugo, who intended to show some samples of these Selous stones to Mary Leakey. In a letter from the Serengeti six weeks later, Hugo reported that he had visited Mrs. Leakey at Olduvai "to show her the stone tools from the Selous . . . The tools are made of quartzite and chert. The ones we found together are from the Middle Stone Age and consist of disc or tortoise cores, and flakes struck off when making similar cores. Dating is impossible since the Middle Stone Age culture apparently covered a wide span of time which started about 200,000 years ago. I did find three handaxes after you left, and these are older. They are from the Acheulian handaxe culture which came to an end 200,000 years ago. Again, this culture covered a large expanse of time so these handaxes could be considerably older."

Later that day, we had a strange sense of timelessness when in a stream bed we found more hominid traces: catfish bones lay on a large flat rock in the stream, beside them the three stones of a small cooking fire. Near the stones lay a long tweezer carved from a green stick, the tips of which had been bound with fiber to hold the fish while it was broiled over the fire. Stuck into a bush, as if the wanderer meant to return, was the traditional flat-bladed wood spoon used for stirring porridge, and a place in the stream bed had been cleared where a man might sleep. We couldn't believe that a poacher would be all alone – for we both had the distinct feeling that this camp had been used by just one person – or would make his camp within sight of the track, although these tracks south of the Liwale–Madaba road were rarely traveled. Who was it, then, who walked by himself in the remote southern Selous, a hundred miles or more from the nearest hut? Perhaps a honey poacher? Whoever it was must have carried a pot and something to cook in it, and he must have been confident in the bush, since there was no sign of a bonfire to scare off animals. Excepting the shadow of his bed and the rude implements at the faint hearth, the unknown traveler had left little more behind than the stone tool makers of hundreds of centuries before him.

Off the track ahead, a kudu cantered back into the scrub before stopping in a screen of thin gray saplings to inspect us; she seemed to know how well these saplings blended with the thin silver stripes on her pretty flanks. Certainly she would never be seen by passers-by in these sere woodlands of the dry season, and I wondered how many of these creatures watched us pass. Then leaf shadows shifted, a huge gray kudu bull was there beside her, big pink ears twitching with apprehension in the soft autumnal light, and through binoculars I could make out a second female, then a third, a few steps back in the brown sun. Hugo tried to come up closer, but the animals overcame their curiosity and withdrew into the wood.

One day Hugo, Rick, Karen, Maria, and I walked perhaps ten miles up the Luwegu until we found a place we liked on a black granite outcrop that rose from a rock ledge overlooking an harmonious bend in the river. In a simple fly camp without tents, we spent two quiet days watching the river, fishing and bathing, observing the local hippo herd, and walking out in the cooler hours to see what we might see. Here, flaked implements of early man could be found on almost any outcropping of rock, and petrified wood, in shards and sections and whole mineralized logs, was littered along the elephant paths and through the grass. At dark, after a simple supper and a bottle of wine, we listened to the night sounds and stars and the soft murmur of the river, and at daybreak came a strange ancient chorus made in the simultaneous needs of frogs and scops owls, hippos and ground hornbills, counterpointed by the deep tearing coughs of a restless leopard.

On this brief safari, we were escorted by Bakiri Mnungu and three young Ngindo. Years ago Bakiri Mnungu had been one of Bwana Niki's porters; he was later made game scout, then a head game scout, a title that had now been changed to "game assistant", perhaps because there was no scouting any more. From his first days with the Game Department until Brian departed in 1973, he had worked under Bwana Niki, and he was proud of this: "He taught me everything," Bakiri said. (Bakiri also remembered Ionides, who, he told us, invariably wore long pants with many patches and a hat of many colors, also patched, and never went anywhere without a walking stick: because the Old Bwana was known to be wealthy, the Africans could never understand why he dressed like a poor man, Bakiri said.) Bakiri is a small, friendly, humorous man who is almost always cheerful, even under stress; one day when shouted at by Philip Nicholson, he merely laughed at him and said, "You're just a young boy, you don't know how to curse an old bird like me." Like all of the Africans, Bakiri understands that Philip, who is young and excitable, has somehow gone astray between cultures and customs and generations in the new Africa; the very fact that he feels free to dispute with them points up the fact that Philip is socially involved with Africans in a way that previous generations were not, and he spends a lot of time at the kitchen fire.

Leaving Mkangira, Bakiri had been extremely upset that due to his own miscalculation about personnel, he was asked to carry a load and not a gun. In a fine tantrum, he cried out that the news of a game assistant carrying a porter's load would ruin his name for ever at Kingupira. However, he shortly got over this, and said so – "I am over it" – picked up the load and then took the gunbearer's place at the head of the procession, although Rick and I were carrying the guns. By the end of the day, in giving orders through Bakiri, Rick had patched up the game assistant's authority, which Bakiri himself emphasized that evening by disdaining to speak to these young Africans of his own tribe who had dared to laugh at him; instead he marched up and down the bank, guarding black and white alike from attack by savage beasts. Next day, on a walk further upriver with Maria, I permitted Bakiri to come along to guard us, and as usual he was lively and helpful, despite a tendency – perhaps left over from the nervous days with Bwana Niki – to load up his gun rather too quickly when elephants appeared. Bakiri said that Bwana Niki was a real *fundi* (expert) when it came to elephant hunting. "Bwana Niki would go so close, he would brush the ear back before he pulled the trigger," Bakiri said, laughing aloud in the memory of his own nervousness, about which he is not in the least sheepish; on the contrary, he declared that he always did his best to stay as far behind Bwana Niki as possible.

"It is terrible since Bwana Niki left!" Bakiri suddenly cried, with a yelp of agonized laughter. "Look at my clothes, nothing but holes! He used to issue us new uniforms, we were very smart, and now I have to hold the ammunition in my hand because all my pockets are broken!" Bakiri plucked sadly at the ragged clothes that were so unseemly in a game assistant. "Every day I was happy because we went here, there, everywhere. Now we don't go anywhere, and we get hardly any money, so I can't retire. I had a big belly – now look at me! I call him Baba – my father – because he was a father to me. It is *very* bad here since he left."

By this time Bakiri was his old self again, curious and enthusiastic, helping Rick by weaving fish line out of ancient rope, entertaining the bored porters, and standing guard with his gun in the shade of a big ebony while we lolled in the warm shallows of the river. The day before a young crocodile had surfaced suddenly at Rick's feet while he was fishing from a rock, and Bakiri declared that from now on, this ledge camp would be known to local Africans as The Place Europeans Were Threatened by Crocodiles and Then Went Swimming.

Brian had described how, on the Ruaha River, Bakiri Mnungu had once warned him about an approaching hippo when he was standing on a log at the riverside. "I looked up and saw this little thing coming downstream toward me, just the snout, and I thought it was a hippo, too, but when it submerged, I suddenly got this funny feeling and jumped back off that log. Two seconds later, this big croc surfaced right where I had been, and not finding me there, disappeared again. So old Bakiri prevented a terrible loss to the world." When I asked Bakiri if he remembered this episode, he laughed and nodded his head, but claimed no credit: Bwana Niki, as he recalled, had been sitting on a log beside the river, and when the crocodile had surfaced right in front of him, the Bwana had run.

Next morning a leopard was seen by Bakiri and one of the Ngindo – despite all the leopard sound and sign, the only one that was actually sighted during our expedition to the Selous. Plains game and elephant and buffalo came to the glades in a large grove of borassus, not far upriver; on two occasions we saw kudu from our ledge, coming down to drink from the brown flood. Here was a chance to move slowly and quietly in awareness of the small sounds and smells, the rocks, trees, insects and small vertebrates – a small and smelly turtle, interfered with in its doughty course along the bank, and also the great river turtles from which the sun glanced as they slid into the currents; the odor of a certain mint, a cat-piss stink so strong that, meeting it suddenly in a close place, I actually stopped short and had a look around; the sick sweet smell of shiny *Strychnos* fruit and the fragrance of wild jasmine and caper blossoms and gardenia; the notes of color in a yellow hyacinth, red indigofera, blue commelina. We learned to listen for the sharp snap of sterculia pods, twisting wide and broadcasting their bronze shiny beans as they sprang free of the tall white trees on the high bank. Maria and I made a collection of the beautiful red-black "lucky beans" of the pod mahogany, and Hugo discovered a miraculous false flower made by a cluster of gaudy tree hoppers, red, turquoise, saffron, and white: the unopened white buds along the plant stem taken over by this lively flower were the insect's larvae, covered in a fuzz of long white hairs, and under the leaves of the tree above, hundreds more of these buds were shedding the white hairs of the pupa stage, as a spray of colorful blossoms came to life.

On the day of our return to Mkangira, Bakiri Mnungu pointed out the big tamarind with its dense shade and thick horizontal limbs where he had seen the leopard the day before, showing us exactly how the *chui mkubwa*, the Big One, had shot down the tree, hitting the ground just here – *hapa!* – and whirling off in a yellow blur into the thicket. Two days before we had seen Sykes monkeys in this place, which may have been its attraction for the leopard, and today we discovered what we thought at first was a dead hippopotamus, embedded in a shallow stagnant pool behind the river. The poor creature lay immobile in thick muck not deep enough to protect it from the sun, its whole shoulder and neck laid wide by a massive tear, and its broad back lacerated by a skein of claw marks; its eyes looked glazed as, very slowly, it turned its head to stare in our direction. While we deliberated whether or not it should be shot, it clambered unsteadily out of the water into the darkness of the river thicket where, according to Bakiri, it fell down. Reporting this, Bakiri laughed, not out of callousness but in that nervous release that is brought on by grotesque events. And we were depressed, wishing we had put the poor animal out of its misery, for in leaving the pool it had revealed an even more hideous wound in its right hind quarter, a great loose mouth of discolored flesh so rotted out with putrefaction that the tissues had all fallen apart and a rush of water came sluicing out, leaving an awful stink of death in the heavy air. We stared helplessly at the dark cave of branches where the hippo had disappeared; there was no sound.

Brian had said that lion killed hippo regularly along the Kilombero, although he could not figure out how it was done; in his opinion, a lion's teeth were simply not long enough to bite through a hippo's heavy hide and still penetrate deeply enough to kill it, and he suspected that the creature must die of shock or heart failure, as the green pigeon is said to do sometimes at the sound of gunfire.

Not far downstream, the death smell came again, this time from a hippo's carcass, swollen a pale purple, that was stranded like a huge rubber toy on a hidden bar out in mid-river. Downstream of the hippo was the dark green-gray head of a large crocodile, and the head left a wake in the brown current as it drew close to the carcass. The crocodile was in no hurry, and probably it had already tried to open up a hole in the tough hide and was now waiting for the hippopotamus to soften. When it lifted its spiny back and tail out of the water, we could see that it was ten or twelve feet long, and now the big head was elevated, too, with the teeth protruding from the long saurian smile; the smile begins in a small loop beneath the brow knob that contains the crocodile's modest brain and stony eyes. Nearing the carcass, it sank from view to take a bite, then surfaced again, approaching the white pasty throat, as a Goliath heron and a palm-nut vulture crossed the foul air, oblivious, and trumpeter hornbills came to the wild date palms by the river, and a pied wagtail tipped and chirruped on a river log greased with primordial mud. Other crocodiles rose and sank away, and all but the first were the off-yellow color of the froth that floats along these rivers; the snouts and eyes appeared, like branch tips, then withdrew again.

Upriver, the mangled hippo had made its way into the water. Immediately it found itself challenged by a bull from a nearby herd, which came for it almost submerged, in ominous silence. The day before, I had seen a hippo flee the water and take refuge in the bush of the far bank when the dominant male came at it from the nearest herd, but this one was too weak to retreat; it merely backed a little, groaned a little, and its antagonist, perhaps detecting from its smell that it was no threat, did not bother to attack the dying creature.

Although we have seen a number of scarred hippos, most of them losers in the constant fighting among bulls, it seemed odd to find a dying and a dead one so close together; perhaps both of them were casualties of the same fight. At the junction of the Luwegu with the Mbarangandu, another dead hippo lay in the shallows by an ancient tree that jutted a long serpentine head out of the flood like a great python. Next morning, this hippo was soft enough to eat, and a whole squadron of spleen-yellow crocodiles floated downstream of it, the wakes of their lumpy heads running together in the river current. A bigger croc had hauled itself out on to the pale hulk, staring blindly into nothingness with every appearance of well-fed satisfaction, and another large individual was feeding, sinking beneath to seize and roll and twist off chunks of the flaccid carcass, then thrusting its long jaws into the balmy river so that the morsel could fall down its throat to be gulped and swallowed.

For several nights, a large six-foot shining black snake – the white-lipped cobra – appeared in the vicinity of Mzee Nzui's cooking fire, a long neat pyramid of ash which rose from the ground as the days went by. Mzee

Nzui demonstrated how, on its fourth appearance, the snake had slid out from beneath the big tin storage chest in which he baked his bread (the tin chest was placed on moderately hot embers, with more embers heaped on top), and spat venom on the arm of one of the young Ngindo. This time the cobra refused to retreat, and was killed with a shovel. However, Mzee Nzui was still nervous. He had never seen a snake behave that way, returning again and again to a crowded place, and with such boldness, and wondered if it might not have a mate. Teased about his nervousness by Rick and Karen, he laughed with them, but not for very long. "I am an old man with two wives and eighteen children," he said, "but I wish to see the world to its finish!"

One night after supper, an *ngoma* or dance was presented by the staff to express their approval of the safari. This traditional event, a gesture of hospitality and greeting and anticipated farewell, is of somewhat suspect spontaneity, since the singers and dancers are paid servants of the audience. But Rick Bonham declared that the *ngoma* had been the Africans' idea, he had not suggested it, and anyway, it was quite clear that staff morale had been high right from the start, a tribute not only to Karen and Rick but to people like Maria, who had interested herself in all the staff activities, and talked and laughed with them, and shared with them the visitors' ideas of what we hoped would be accomplished by a safari that was their safari, too.

The dancers dispensed with Western dress in favor of bare torsos and skirt-like kangas, and because their songs and dances differed, as did their use of Swahili, the Kenyans danced first as a group and the Tanzanians second. The Kenyans were led by Mwakupaulu, the assistant cook, who took the woman's part by tying up his kanga around his neck in *bibi*-style and using pleats to achieve an effect of big loose hips and breasts. Mwakupaulu kept time with a rattle made from a coffee tin filled with pebbles, while Kazungu played a tom-tom drum fashioned that afternoon from impala hide and hollowed wood, and John Matano, the truck driver, played a sort of triangle made by striking two wrenches together. The lead singer was Charles Mdedo, the versatile mechanic who also helped in the mess tent and the kitchen, and Mzee Nzui lent dignity to the business by shuffling mightily back and forth and pounding the rhythm into the earth with a staff longer than himself that made him look like some ancient prophet from the desert. The one Kenyan dancer who did not have his heart in the *ngoma* was Kirubai, who moved vaguely back and forth, avoiding the gaze of the pleased audience by staring straight upward into the starry night, as if studying bats. It was assumed that Kirubai felt shy, self-conscious, but even though there was good feeling in this camp, I had to wonder if the *ngoma* might not strike a hunter of elephants as undignified, even demeaning.

At the start, Charles Mdedo made a brief warm speech of welcome, saying that the staff had been very happy on this safari and wished to sing their guests a few songs to make us happy, too. The first song was a song of celebration of the happy safari, the second was a dancing song that would ensure our safe return journey, and the third was a hunter's song (nothing at all like the hunter's song that Kirubai had sung out in the bush) in which, having courteously warned the audience not to be surprised or afraid of what was coming, the Kenyans came forward and picked up the apprehensive David Paterson in his camp chair and carried him high around the fire, in celebration of the tasty buffalo he had helped to shoot. David, a bright and energetic person who had been very good company on this safari, was as enthusiastic about the hunter's dance as he had been about the hunting, and carried things off with high spirits and good-humored shouting.

The Tanzanian part of the *ngoma* was commenced by Mzee Saidi, who told the audience how honored he and his people had been by our kind visit, and how grateful they were to Bwana Niki for having brought us: the guests were coming here, and too soon they were going, and these dances and songs would express the sincere thanks of all the staff. Being younger, the Tanzanian dancers were most lively. The first dance was a traditional dance of the Ngindo, the second was a dance of thanks made up for the occasion, and the third was *Kwaheri nenda Salaama*, Goodbye, Have a Safe Journey Home.

Now Goa came forward and repeated Saidi's sentiments in his deep, shy voice, staring intensely into the face of Brian Nicholson; he said he was sorry that what Bwana Niki had accomplished here with the Game Department had gone all to pieces, but he wished us a safe journey, and hoped we would come back again. Brian made no speech of acknowledgement, but afterward, without drawing attention to himself, he went over and thanked the staff for the *ngoma*. "It is nice to be back here and work with you again; you've been doing a good job, and I want to thank you," he told them. "And now I've asked you to do something difficult, as in the old days. In a few days some of you will walk out with loads on your heads on a real foot safari: it's good to know there are still some real men left in Tanzania!" The Africans laughed, very pleased with the whole evening, and not less so when another case of beer was ordered with which they might finish the *ngoma*.

One day down near the airstrip by the Mbarangandu, a lone bull elephant was seen wandering slowly back and forth along the river, as if it had lost its last sense of direction; when Hugo approached, it actually drew near the car – "but not at all in an aggressive way," he said. "It had what looked like a spear wound on the side of its face, and was holding its ears back tight against its neck in a strange manner, and it seemed to me that its jaw was swollen. Then it wandered away again to a small pool where it sprayed a little water on its head and beneath its ears."

That night or early the next day, the elephant sagged down and died against the green grass bank between the plain and the white sand of the river, and a day later, more than three hundred vultures had assembled, including one huge lappet-face and a few white-headed vultures, which we had not seen before in the Selous, and even two beautiful palm-nut vultures, which may have joined the madding throng for social purposes, since they are not known to consume carrion. The first to arrive shared the carcass with hyena and lion, but perhaps these animals were already well-fed, for as the hordes of dark birds circled down out of the sky the carnivores withdrew, and the elephant disappeared beneath a flopping mass of vultures that stained the river sands all around a dark gray-green.

The elephant carcass was inspected before the birds reduced it to a cave of bones, and as it turned out, the left ear it had held so close was protecting a great infected wound that maggots had eaten out down to the bone; there was also the separate wound on the left side of the head that looked as if it might have been made by a spear. The game scouts say that local poachers of the region don't use spears, only arrows tipped with *Akokanthera* poison and old musket-loaders armed with poisoned shot; they thought that the larger hole looked like a musket wound, and this opinion, which Brian Nicholson endorsed, was lent support when Philip heard a shot back in the hills. Not that the two episodes were related; the elephant's putrefying wound was some weeks old.

But Kirubai thought that the two wounds were caused by tusks of another elephant; sometimes, he said, bull elephants will fight so violently that tusks are broken, and an elephant may wander around for months with such a wound before it dies. As a former "poacher", he had not seen much evidence of poaching, at least not here in the far south of the Selous; he doubted very much that poachers would have overlooked the five valuable tusks that had been found while we were here. (Alan Rodgers estimates that if ivory collecting were efficient, the Reserve might produce some twenty million shillings worth each year from tusks of animals that die naturally.) With the tail hairs Kirubai made a traditional bracelet for Karen Ross, and although Karen dislikes ornaments made from animal parts she will keep this bracelet because it was made by Kirubai at Mkangira.

Though Bakiri Mnungu went along with Bwana Niki's view that this elephant was a victim of the poachers, he also agreed with Kirubai that there was no real evidence of poaching in this part of the Selous; on the other hand, he had seen too many dead elephants. No, elephant numbers had not been much reduced, in his opinion, but on the other hand, nobody had been out to look at them in recent years, so who could say? In the old days, there was none of this sitting around at the game post, the scouts were off on safari all the time, sometimes for two or three

months, getting to know the bush. There was no poaching then to speak
about, Bakiri said: we reported the presence and abundance of game, we
made note of water locations and put up signs, and we made fires in the
first part of the dry season. All the roads and all the vehicles were in good
shape, and because of the regular burning of long grass, the animals could
be seen and counted, and everybody knew just what was what. Now
everything was *kufa* – dead – and nobody went anywhere or looked at
anything.

Not long after the death of the elephant, five game scouts turned up
at Mkangira, having made a five-day walk in from Liwale; their
instructions were to rebuild the fallen game post, then await a hunting
safari that was supposed to enter this region some time in October. Brian
Nicholson just shook his head, perhaps suspecting that this show of
efficiency had been put on for his benefit, or that the Game Department
wished to keep an eye on us; he spoke sarcastically of these brave game
scouts performing their daily patrol between their cooking fire and our
kitchen, and one day, as the scouts – who were outside – just stared
at him, he made their polite welcome an excuse to enter their small
stockade without permission, as if carrying out an official staff inspection.

When the airstrip repairs were completed, Bakiri Mnungu did not
have much to do, and spent a lot of his time out at the game post in the
company of these Ngindo game scouts. Since our walk up the Luwegu,
Bakiri had been very friendly, bringing fresh tamarind pods to our tent
because he knew that we liked the astringent taste, and one morning
as I walked past the post on my way to the north plain beside the river,
Bakiri called out to me to come and join the scouts in their breakfast of
uli gruel, or "*porrigi*". Because I had eaten, but mostly because of my poor
Swahili, which I thought would make our conversation painful, I
declined, continuing on my way with a grin and a wave. Immediately I
felt vaguely depressed. Probably I had been impolite, and also selfish; the
exchange that would have been painful for me might well have been
entertaining for the game scouts, who were always amused and
encouraging about my Swahili and in any case would have carried me
along out of kindness and courtesy. Going on my way alone, with my
white man's private notebook and binoculars, I knew I had missed a
warm and vital chance.

From my place beneath a big tamarind near the Mbarangandu, on
what I thought of as the "northern plain", I could see for several miles up
both the rivers. Where the currents met, the waterbuck moved out across
the sand, the bucks sparring half-heartedly as the does, paying no
attention, moved on past. The other evening, on a social impulse, the
solitary wildebeest that lives on this northern plain joined the waterbuck
as they galloped toward the river. The heavy waterbuck drag their hooves
as they run, and they kicked up the white sand in a fine display, while the

wildebeest, bringing up the rear, merely rocked along with a spirited whisking of its long tail. Crossing the shallows, the waterbuck sent the spray flying, but the wildebeest, feeling water beneath its feet, ricocheted off with the weird kick and buck peculiar to this species, then high-tailed back to the dry ground on its own side of the river. There it resumed its solitary life, which risks the attention of the lions, since solitude in a social species is often sign of illness or decrepitude. But perhaps the lion were distracted by the wart hog and kongoni on the plain, which were followed about these days by their new young.

By early September, in the diminishing rivers of the dry season, all but three hippos that resided in a deep pool under the south bank had abandoned the shallow Mbarangandu for the Luwegu. At this season on the Luwegu, a clump of hippo heads broke the bland surface of the river every half mile, with two herds very close to the river junction, and there was no time in the day when one did not hear them. The waterbuck also seemed to like the confluence of rivers, where the broad sand bars afforded them safety from the lions, and numbers of water birds came there, too, including a big flock of African skimmers. When not flying up and down with their long lower bills stuck in the water, eyes focused on the little fountains they create for the small fish and other creatures that they live on, the skimmers sometimes soared in pairs in wonderful courtship gyrations on the blue sky, or mobbed the kites and herons that dared to fly across their delta. They seemed to live peacefully enough with kingfishers, and also with the sandpipers and plovers. In recent days, the African shorebirds had been joined by Palearctic migrants from Eurasia: greenshanks and the little stint, and the marsh, green, and curlew sandpipers descended from night skies to rest on these warm margins.

Standing barefoot on white African sands, smelling the damp algal smells, the mineral rot of driftwood, I studied the tracks of hyena and lion, hippo and elephant, the foot-dragged prints of waterbuck, the ancient hand-prints and serpentine tail furrow made by crocodile, to name just those I could see from where I stood. The air was filled with engaging dung smells and the protest of hippos and "yowp" of monkeys from the trees across the river.

Although buffalo and elephant were here at Mkangira when we arrived and on the second day a rhino was seen by the Nicholson family on an outing after tea, these large beasts soon vanished from the region; unlike parks animals, they avoided the presence of man. But after the middle of September, as if anticipating our departure, a herd of several hundred buffalo came down to the south bank of the Mbarangandu, and the next day two rhinos appeared in the same place; elephants reappeared on the northern plain, and a loose herd of eight bulls, including one with a large single tusk, could be seen each day out near the repaired airstrip. Discovering the dead one by the river, these elephants stayed near it for a day or two in answer to some elephantine instinct, perhaps more akin to respect for death than man chooses to think, although the dead kinsman was now no more than a hollow gray mound of hard-baked skin, a sagging armature of bone.

Almost every night restless lions could be heard on both sides of the river, and sometimes leopard, and invariably hyena; because of the smell of the buffalo and impala that were killed every few days to feed the camp, the hyenas were bold nightly visitors, skulking about the kitchen area and between the tents, leaving behind the strange long prints that like the rest of their appearance is more suggestive of the dog than of the aberrant cat that they really are. One night another hyena clan made its own kill on the far side of the Luwegu, filling the night with excited whoopings that turned to high eerie giggling and laughter. Out there in the dark where the hyenas were tearing the wide-eyed victim into pieces, those crazy noises would be ringing in its ears.

The brown flood sparkling under the moon was perhaps two hundred yards across, yet it was shallow enough for a man to wade the chest-high water were it not for the big crocodile that had showed itself now and again in recent days and took most of the languor out of bathing. Between bird calls, in every silence, came the soft wash of the two rivers, pouring away to the north and west to meet the great Kilombero that comes down out of the Nyasa Highlands on its way to the Rufiji and the sea.

(Above) Petrified wood. (Right) Boomslang.

IX

On an early African morning, Brian and I set off for the south, wading across the Mbarangandu not far above its confluence with the Luwegu. Climbing the ridges between rivers, we shall follow the game trails for about eight miles, then descend to the Luwegu and continue south for perhaps three days before turning east to explore a high plateau with its own extensive swamp or pan. From the plateau, we shall descend a tributary river that comes down off the west escarpment of the plateau and turns northeast, arriving eventually at the Mbarangandu. There is no good map of this region, and neither the plateau nor the river has a name; as for the pan, Brian Nicholson is the only white man who has ever seen it. Excepting Ionides and Alan Rees, the warden of the western Selous, he is the only white who has ever walked through the vast southern reaches of the Reserve. "This is the heart of the Selous," Brian told me, "and you and I will be the first into most of the country that we're going to cross." If the water has subsided enough in the next fortnight to permit Land Rovers to come upriver, bringing supplies, we shall explore still further south, up the Mbarangandu; otherwise we shall head back downriver, returning to Mkangira in about ten days.

All but the Nicholson family, still asleep at camp, and David Paterson, who is sick with fever, have come to see us off, and Maria and Hugo wade across the river with us under the armed escort of Bakiri Mnungu. When Hugo tells our porters not to notice his camera, there is an outburst of laughter over the fact that Africans, too, are to be photographed, and one porter's sweet squeal is so infectious that in a moment all of us are hooting senselessly, to let off the nervous energy of the departure. Bakiri Mnungu makes the young Ngindo laugh harder and harder, watching the white man as he does so, and Renatus takes such delight in this convivial moment that he is literally falling about on the white sand. Even Nicholson is grinning. Then he says gruffly, "*Haya tayari, basi twende!*" "Let's go!" Looking worried now, Renatus calls goodbye to his friend Abdallah, the young porter with the infectious laugh, and Abdallah calls back, "*Mungu aki penda, tuta onana!*" "If God pleases, I shall see you again!"

At this place the river is edged by high elephant grass where big animals might be hidden, and as we pass into it and our friends disappear, the laughing young porters fall silent in an instant, as if entering the unknown. For the next hour, as the sun rises and the file of men climbs to the open woodland of the ridge, there is no sound but the tentative duets of barbets and boubous, and the soft whisper of our passage through dry grass.

In crossing the Mbarangandu, leaving behind the tracks and Land Rovers, the tin-roofed game post, the green tents of our camp at Mkangira, we have also left behind all roads, all sign of man, and in doing so, we seem to have entered a new Africa, or rather, "the Old Africa", as Brian calls it: behind the heat and the still trees resounds the ringing that I hear when I am watched by something that I cannot see. "You're getting the feel of it now," he mutters, peering about him, for he, too, has sensed the power and the waiting in the air. "Only people to come this way in years, I reckon; I don't think the bloody poachers have got this far." Years ago, he had laid out a track for patrolling this part of the Mbarangandu, but all that was left was the pathway made by the round wrinkled pads of elephants, in the silent years without sound or smell of man when the huge gray apparitions had followed the abandoned road. The shadow of the road is only visible to the eyes of Goa, and soon it vanishes in the sun and dust.

As tracker and gunbearer, Goa is in the lead, a rifle over his small shoulder with the butt extended toward Brian Nicholson. The gun is a heavy-bore .458 of Belgian make, an "elephant gun", very useful for stopping large charging animals. Goa holds his free hand far out in front of him, as if extending it to be kissed, fingers pointed down as if to dowse the ground before him for the slightest sign or sound or scent of danger; he moves so lightly that he seems to rise ever so slightly off the ground, at the same time craning his head as if to see over tall grasses that, much of the time, are well above his head.

There are six porters, young Ngindo who were recruited from Ngarambe village, just outside the eastern boundary at Kingupira, and behind the porters, making sure that none falls by the wayside, is the young Giriama camp assistant named Kazungu, who will serve as cook.

Kazungu did not wish to accompany this foot safari because he thought he would have to carry a load upon his head, like these unsophisticated young Tanzanians; as late as yesterday, he was complaining of an excruciating pain in his right foot, screwing up his lively face for emphasis. But when informed that he would only have to carry his own gear, together with a panga for cutting firewood and brush, he was happy enough to come along; in fact, as I discovered later, he kept an enthusiastic journal of the safari which he and his friend, the Taita mechanic John Matano, translated from Swahili into English and were kind enough to let me use:

> We began our safari at the junction of two rivers, the Luwegu and the Mbarangandu, and the date that I left was the 2nd September 1979. We were eight of us and two Europeans, one as our guide, whose name was Bwana Niki, and the other a book writer from America whose name is Mister Peter. And I was the tenth one, as the cook. One of us was an askari of the bush [the game scout, Goa] so we had no doubt with wild animals. We walked for a number of kilometers until we came to the Luwegu ...

The porters take turns carrying the old-fashioned tent that Brian wished to bring along, despite my feeling that we did not need it. Since I am carrying nothing but binoculars and notebooks, I feel slightly ashamed, whereas Brian is not sheepish in the least. "If I had to carry one of those loads in this sort of heat," he admits cheerfully, "I wouldn't last out the first hour. I've never carried a thing on trek in forty years, and I never shall." What I was hearing wasn't laziness – Nicholson is anything but lazy – but a principle left over from the reign of Ionides, who liked to say, "I never do anything that can be done for me by somebody else."[1] Brian is proud that all his old safaris were elaborate – far more elaborate, as he says, than this one. "Always had my own tent, of course, with tent fly and camp table and chair, and my gunbearer and cook and a hell of a lot of porters. Sometimes I'd be out five months at a time, so I needed a lot of equipment, but also it was important to be comfortable. Took along whatever I wanted, as a matter of fact. I had one man who just carried books, another who carried a coin chest for buying food in the settlements; the rest carried my personal gear and the food for all the others. On short safaris through settled areas, I had fifteen porters; on long ones through the bush, I would have forty. But once the food started to go, I couldn't have all those people sitting around eating up what was left, so I'd send them back in lots of six; they couldn't be sent off in ones and twos, not in *this* country."

On the ridge between the rivers, the file moves rapidly, in ant-like silence, as if in flight from the accumulating heat. Since leaving the Mbarangandu, we have encountered no animals at all, only the pale rump of a kongoni, vanishing like a ghostly face into high grass. Other animals have come and gone – we see a rhino scrape, the elegant prints of greater kudu, old droppings of elephant and buffalo – but as the day grows, so does the sense of emptiness in the still woodland, which is not a closed canopy but open to the sun, and everywhere overgrown with high bronze grass. "All dead, dry stuff," Nicholson mutters. "No good to animals at all. In my time, the whole Reserve would be burned over every year, two at the most; I had over four hundred game scouts who spent most of their time out in the field, and burning was their main job." He murmurs to Goa in Swahili, then stalks on, and Goa steps off the elephant path and sets fire to the tinder grass, which ignites with a hollow rush of the dry air. The fire leaps up with a hungry crackling, and a dark pall of smoke rises in our wake as we move southward.

This thin, tall man walking ahead of me in his big floppy hat, old shirt and shorts, and worn red sneakers looks more like old Iodine must have looked than the conventional idea of the East African professional hunter, or the crisp old-style warden in regimental khaki: I like this "Mister Meat" for his lack of vanity. In his angular, stoop-shouldered gait, he keeps up a long easy pace, remaining close to the swift, effortless Goa, yet every little while he turns and casts a hard, bald eye back along the line, noticing quickly when the porters fail to keep close ranks in river thickets and karongas, or when one or more tends to fall too far behind. "*Wanakuja?*" he calls. "Are they coming?" And with the barrel tips of the shotgun that he carries he moves a thorn branch off the thin trail, anticipating the bare feet of the young porters. His concern is professional – foot injuries will cripple our safari – but it isn't unfeeling, whatever he might have one believe.

Ahead, three young bull elephants are standing beneath a large and dark muyombo tree, which at this season, in anticipation of the rains, is covered with a red canopy of seed pods. Getting our scent, the elephants move away in no great haste as we come down into a grassy open glade. The blue acanth flowers of dry ground give way to blue commelina and lavender morning glory, and there are meadow springs and frogs and singing scrub robins. "At this season, most *miombo* is pretty dry from one end of Africa to the other, but here in the Selous it's so well-watered that these little paradises occur everywhere in the dry woods," Nicholson says, as the porters set down their loads beneath a tree. "That's why we didn't bother to bring water bottles." But Goa is out putting the torch to the dry grass all around, and over this paradise black smoke is rising; within minutes, the racquet-tailed rollers appear, filling the crackling heat with strident cries as they hawk the insects that whir up before the flames.

We head southwest across the river bends to the Luwegu. Unlike the Mbarangandu, the Luwegu still carries a swift flood of brown-gray water that in most places fills the river bed from one side to the other. This high water, unusual at this time of the year, must account in part for the scarcity of animals along the margins, since there is more water than they need in the pools and springs back in the woods. Where we come out on the banks of the Luwegu we see no elephant at all, only a large crocodile which lies out on a bar along the bank, its jaws transfixed in the strangled gape with which these animals confront their universe. In the mile between bends of the river two large herds of hippopotami are visible; it seems likely that there are too many, that one of these long, slow years there will come a great dying-off of the huge water pigs, to bring their numbers back into balance with the wild pastures that they have pushed further and further from the banks. According to Brian, such dying-off occurs in the Selous about once every seven years, in separate places; he remembers it once in the Ruaha, and another time on the Kilombero. But today they steam and puff and honk in great contentment, though two get at each other every little while in a great blare and thrash, to banish the monotony of river life.

Everywhere as we walk upriver the animals are starting to appear; it seems to be true, as Brian claims, that here in the Selous the animals are not especially active early in the day, as they are elsewhere, and do not move about until mid-morning, though *why* this might be true is not clear. Among the smaller animals that cross our path are ground squirrels and the green monitor lizards, small relatives of the great Komodo "dragons" of the East Indies, and a black-tipped mongoose, scampering along the bank, that is red as fire; the banded and pygmy mongoose are common in the Selous but this is the first of this weasel-like species I have seen. Impala are numerous and remarkably tame, and a band of waterbuck under a tamarind beside the river lets man walk up within a few yards before prancing off in a pretty canter into the woods; further north these animals would take off at a dead run at the sight of vehicles, which ordinarily disturb them less than a man on foot. Wart hog and wildebeest are also rather tame, though not confiding. Under a big tree by the corner of the river, from where the Mahoko Mountains can be seen off to the west, stands a placid group of elephants; not until we move a little uphill to the east of them, to let the breeze carry down our scent, do they give way. Kazungu described the scene in his journal:

> We saw elephants where we wanted to pass. We went upwind of them to give them our smell, and this make me understand that no matter how dangerous an animal is, if he is not familiar with a smell, he will run.
>
> We went up and down the hills and met some different animals.

Behind the elephants is a large grove of borassus palms, with their graceful swellings high up on the pale boles; from each palm, or so it seems, a pair of huge griffons violently depart, their heavy wing beats buffeting the clack of wind-tossed fronds. At this season, the borassus carries strings of fruit like orange coconuts, which are sought out by the elephants; here and there in the dry hills, far from the nearest palms, lie piles of dried gray borassus kernels, digested and deposited, from which the last loose dung has blown away. The mango-like kernels remind Brian of the elephant habit of gorging on the fallen fruits of the marula tree. "Used to ferment in their stomachs, make them drunk or sleepy; they'd just lie down on their sides and snore. Ever hear an elephant snore? Oh, you can hear that a *hell* of a long way!"

Keeping the porters close, we push through thickets to a shady grove beside the river. This first day we shall quit early, while the sun is high, to give the Africans a chance to dry the strings of dark red meat jerked from the buffalo killed yesterday near Mkangira; the biltong will be their main food for the journey. Goa and Kazungu string a line between high bushes on the bank to hang the meat, which is brought up in big handfuls by the porters; once dry, the biltong is very hard and tough and may be slung around amongst the luggage.

Abdallah spreads green canvas in the shade, and the sahibs lie down upon it to take tea. Since Brian's red sneakers are blistering his toes, the decision to stop early is a good one; also, the Ngindo are not trained porters and will collapse quickly in this heat – it must be 100 °F or more – unless they are given a day or two to get broken in. The remarkable Abdallah of the squint eye and sweet laugh is now doing headstands in the sun – actually jumping on his elbows in a small circle on the thorny ground – but two of his companions are laid out like corpses. One of these is Kalambo, who wears huge blue boxer's shorts with a white stripe that extend below his knobby knees, and the other is a heavy boy with the name "David" who wears a bright red shirt. Then there is Amede – short for Muhammed – who walks with the sway of a giraffe, and Shamu, whose face is wide-eyed, innocent, faintly alarmed, like the face of a small antelope; his small size and slight body, his expression, make him look too young to be carrying loads that might bend his bones, but on closer inspection I see that Shamu is a full-grown man who has retained a child-like air of innocence. Most of the time he sits quietly to one side, smiling delightedly at the witty sallies of his friends.

Finally there is Mata, small-faced as a vervet and given to harsh yapping barks; Mata likes to walk apart from the file of porters, and once or twice he dared walk on ahead when Goa and Brian had paused to light their fires. Bwana Niki had murmured something in Swahili which brought a bad look to Mata's face; he seemed to consider a spry remark, but then thought better of it.

"You'll have trouble with that one," Rick Bonham had said, but the Warden knew better. "Once you put a bit of distance between them and the camp, and they have to depend on you for their protection, they're quite anxious to please," Brian had said, and he was right; after the first encounter with the elephants Mata fell back into line, and gave no more trouble after that first day. "Oh, there are exceptions to the rule of course," Nicholson reflected later. "Once had to kick a porter in the balls and give him a good clout to go with it. No choice, really." When he says these things, old Mister Meat fixes me with that bald eye and gives a faint clack of his false teeth, and I can't make out whether he intends this as a literal account of makeshift discipline in the bush, or if he is teasing my "American" notions about Africans, or whether he is feeling nostalgic about the grand old days when Ionides could have a whole Ngindo village flogged and get away with it. But as it turned out, he meant just what he said: the offending porter, Brian told me later, "was using blackmail in a very remote place, threatening to dump all the loads and abscond with his fellow porters unless I doubled the agreed pay. He was a huge man, who could have torn me apart easily if I hadn't disabled him."

By early afternoon the clouds have gone, and the day is dry and hot. Drinking gratefully from the brown river, I realize how rare now are the places left in Africa where one can drink the water without risking bilharzia or worse; in the Selous, one can sip with impunity from pools and puddles and even from big footprints in the mud. Later I find a safe bathing place behind a silvered log, and lie back for a long time in the warm flood, watching the western sky turn red behind a gigantic baobab across the river. Behind me in the forest, an elephant's stomach rumbles – or perhaps the elephant is pondering my scent, for Brian says that what is usually called stomach-rumbling in elephants is actually a low growl of apprehension and perhaps warning. Trumpeter hornbills gather in the mahoganies over my head, and I am attended by a small dragon-fly, fire red in hue, that might have flown out of the sunset. I am extremely content to be here, yet I do not look forward to the evening; Brian and I have got on well enough, all things considered, but other people have always been around to smooth over the rough edges. I don't know this man as yet. We have been thrown together by fate, not by affinity, and doubtless he regards our enforced companionship as warily as I do.

When I get back to camp the Warden is lying on his cot before his tent, head raised on his elbow, watching me come. I am not surprised that he and I have been having the same thoughts, or when he says, coming straight out with it, "You know, Peter, when this idea came up that you and I should go off together on a long foot safari, I was dead set against it. As late as Kingupira, I was telling Tom Arnold, Absolutely not! You can't just go bashing off into the bush with some fella you've never set eyes on; had to have a good look at you first." Especially, I thought, after having read my book on Africa – and once again, Brian anticipates me. Although

[135]

we have been together for two weeks, he mentions the book for the first time, and actually says, "I thoroughly enjoyed it." Together we laugh about that first night in Dar. "Didn't know *how* to act!" Brian said. "Didn't know what was wanted of me, really."

Not only his words but his whole manner confirm what I thought I had already noticed, that he has left his mask behind at the main camp in Mkangira. He seems happy and relaxed, eager to talk, and the talk is almost entirely free of that cynicism and intolerance that I thought would cause trouble between us. Over our simple supper of beans and rice, he speaks with real affection and respect of some of the Africans on his old staff, such as a scout named James Abdallah who as a youth had been conscripted to help haul that steam engine up there to Madaba. "Terrible work. After three months, I think, James ran away and hid out in the forest."

In those early days, Brian says, he spent most of his time on elephant control, which was often a bit risky. "I lost a very good man once to an elephant, one of my head game scouts, like old Saidi – you might think of a head game scout as a sort of regimental sergeant-major. Today they call them game assistants: I suppose they think it's less demeaning to be an assistant than to be a scout! But anyway, this other Saidi – Saidi Nasora Kibanda – was a first-class shot, which is very rare among Africans; he was also a superb hunter, and very knowledgeable, one of my best men. One day about 1965 Kibanda was out with a trainee on elephant control, and the trainee wounded an elephant which Kibanda had him follow up. I'll never really know what happened; when these things occur, the survivor always puts things in the best light for himself. But apparently the elephant attacked, and the trainee ran, and for some reason, Kibanda failed to stop that elephant, although they were in open country and it should have been a routine shot. That elephant destroyed him." Brian paused as if granting Kibanda a moment of silence. "Old Kibanda was a hell of a good fellow, and his death came as a hell of a shock to me, I don't mind telling you – that man was my right arm. Very loyal, very intelligent – a very nice man altogether."

Brian clears his throat, frowning a little. "I suppose I lost one or two game scouts every year, out on control work, but I can't think what could have happened to Kibanda. These large-bore precision weapons stop an elephant pretty easily, although most Africans have a hard time believing it; perhaps that's why they don't aim properly. Usually the animal will come at you with his head low, and you shoot at the forehead, above the eyes, to hit the brain; if the head is raised, of course, you shoot just at the base of it. Either way he goes right over, no problem at all.

"Of course there are times when nothing seems to go right, and probably it was one of those times that caught up with poor old Kibanda. I had a day like that myself. I was stationed on Mahenge at the time. We had been asked to deal with three bull elephants that were getting into the shambas, then becoming aggressive with whoever tried to drive them out, and running around knocking over houses, too. The people had these flimsy sort of *kilindos*, or huts-on-poles, that they would put up to keep watch on the crops; when elephants came, they would try to drive them off by beating on tins and the like. One day there was an old woman in one of these huts, and when three bulls turned up in the gardens, she started beating on her tins, and the elephants came for her, knocked over the hut and trampled her and tore her to pieces – made a real mess of her.

"When I arrived and finally had them located, I had to stalk them through very high grass, over my head, and when I came up with them, they were all together in a kind of opening they had beaten down in the high grass under a sausage tree. Two were broadside to me, one completely screening the other, and the third was looking off in another direction. When you shoot an elephant in the brain, it always sits back on its haunches, and I reckoned that the near one, falling back that way, would give me a fair shot at the one behind it. Then my gunbearer would give me the other gun, and I'd have two barrels to use on the third animal.

"Well, the first part of the plan worked well enough; the first sat backward, I dropped the second, and grabbed the other gun. But the third animal was already taking off, and I got off two quick shots aimed at the pelvis, because a pelvis shot will cripple an elephant and stop him so that you can finish him off properly. However, I had shot too fast, and he kept going. Because I wasn't absolutely sure the second elephant was dead, I told the gunbearer to load up the first gun, showed him just where he should point the barrel, and then I took off after number three. He was only about sixty yards away, and badly hurt, but my third shot didn't drop him, I still don't know why, just set him running again, and a fourth shot intended for the pelvis didn't stop him either; all it did was turn him right around. I realized that the gun was empty and I had no more cartridges, hadn't thought I'd need them, you see, and I did the only thing I *could* do: I ran like hell back down that path that the elephant had made through the tall grass. After about five yards, I tripped and fell, and he was on me.

"This elephant was badly hurt, and his trunk was full of blood, spraying all over me, but he couldn't smell me, you see, and after he missed me with his tusks, he somehow lost me, and I was able to roll away. Luckily, he decided to take off again, and I ran back to my gunbearer. I didn't have Goa at that time; this man was only a trainee who never did make it as a gunbearer, and when he saw me all covered with blood, he panicked; he thought the elephant had got me, and because I was running, he imagined it was still hot after me, and so he departed in the opposite direction. When I caught up with him, I gave him a hell of a clout to calm him down, and grabbed a handful of cartridges out of his pouch and went back and finished off that elephant. That was the closest call I ever had with a wild animal."

◆ ◆ ◆

[*137*]

A half moon rests in the borassus fronds over our heads, and a tiny bat detaches itself, flits to and fro, and returns into the black frond silhouette. We lie peacefully upon our cots and watch the stars. From the forest comes the hideous squalling of frightened baboons trying to bluff a leopard, or so we suppose, since there is plentiful leopard sign around the camp.

In the moonlight the bull hippos of the herd move in close to the bank to bellow at our fire; in trying to frighten us away, they panic one another and porpoise heavily away over the shallows, causing great waves that carry all the way across the river and slap on to the mud of the far bank. Man does not belong here, and the hippopotami cannot seem to accept us; we have disrupted their whole sense of how the night time world unfolds in the Old Africa. They do not go ashore to feed but remain out there just beyond the fire light, keeping watch on the intruders and banishing our sleep with outraged bellows.

This is how Brian has arranged the day. We shall rise at six, have tea, and be off at seven; at about nine, we are to break for tea and porridge. At mid-morning, we shall walk again for two more hours, then rest in the hot part of the day; in mid-afternoon, we shall walk for two hours more, tending toward the river, where we shall make camp at the first good site.

Leaving the river forest in the early morning, we emerge on to an open plain and head south again over low hills between broad bends of the Luwegu. The bright green grass around a pan has attracted an early morning convocation of impala and wart hog, baboons and geese, and the pan is full of baby crocodiles; storks and herons stand about, minding their own narrow-brained business, and a pair of skimmers, feeding together, draw fine lines up and down on the still surface.

This morning everyone is silent. The porters speak just once, murmuring softly amongst themselves, and instantly Goa stops and turns and, after a moment, says almost inaudibly, "*Nani ani ongea?*" "Who is talking?" When nobody speaks, he takes this for an answer and goes on. Over his shoulder, Brian murmurs, "I always have it as a rule: no one talks on trek except me and the tracker." I nod in approval, not bothering to answer. Talking almost invariably detracts from the real pleasure of walking, in which one finally enters the surroundings. And in the wilderness the human voice is disturbing to animals we might otherwise see, quite apart from the fact that nothing must intrude upon Goa's apprehension of his surroundings.

Now Goa has stopped again, raising one hand. Shadows deep in the scrub ahead have shifted, and soon a bull elephant moves out into the open, in no hurry, since Goa has left him time and space in which to take his leave. There are more elephants during the morning, in twos and threes and fives; although on foot, we are rarely out of sight of them.

A solitary eland bull, a glimpse of kudu. Huge ground hornbills fly away with the slow, ghostly beats of their white primaries that seem incapable of keeping such large birds aloft; like vultures, which are also huge and without enemies, the *batutu* seem exceptionally shy. Yellow hyacinths shine in the grassland, and a bush of daisy-like composites, with a solitary red-stemmed lily unlike any flower we have seen. The country is more open here than it is to the north of Mkangira; the white-crowned black chat of the *miombo* woodland is no longer common, and the racquet-tailed roller has been replaced by the lilac-breasted species of the savanna. Even the tsetse seem to have lost the appetite they show in the closed woods; I watch them alight on the shirt of the man ahead, but they do not bite.

Against the blue hills to the west stands a cow–calf herd of elephant. Getting our scent, a young cow leads the juveniles away while the old matriarch stands guard, trunk high, as if in warning; and soon we see the young toto hastening away after the others, the top of his small earnest head scarcely visible in the high grass. Not far off there is a wildebeest, then a mixed band of impala, waterbuck, and eland; the antelope move calmly to the wood edge, unafraid. For a long time the big gray eland bull stands watching us, attended by three soft brown cows with calves. Like other striped animals – kudu, bushbuck, zebra – eland are taboo animals to the tribesmen to the east of the Reserve, who know that eating striped beasts may bring on leprosy. (This view is shared by tribesmen of the Sudan–Zaire border, who will not eat bongo.)

On the far side of the wood, we find ourselves among a herd of waterbuck almost before a long calm look persuades them that it might be best to take their leave. In numbers of animals seen this morning the far south of the Selous compares with the great parks, but the tameness of almost every creature in this country south of Mkangira has nothing to do with the aplomb of sophisticated parks animals which, being resigned to the human presence, are not tame so much as half-domesticated. Here the confiding curiosity appears to stem from a trusting innocence of man, unlike anything I have seen elsewhere in Africa.

Once again the Luwegu comes into view, set off by broad white sand bars and tall palm trees, the surface broken here and there by clumps of rocks that turn out to be hippos. We descend to the bank at the end of a sand bar where a giant kingfisher, night blue and chestnut, has seized a fish too big for it to manage; it squats on the sand belaboring the fish, then gives this up and lugs its catch away across the river, which at this bend must be three hundred yards across.

As Goa sets his fires we head south, trying to cut across the river bends. Since the river bears west, we are soon inland and higher than we intended, emerging at last on the cliff of an escarpment. Five buffalo lying down under black boulders heave to their feet and hump away into high grass, and from the rocks, a black eagle takes wing across a hidden valley of baobab plains and grassy glades and palm and water pans, cut off from the world by the steep cliffs and hills east of the river. Brian had never known that this place existed; in his years here, he had always found a way around these rough escarpments. "If I'd known about it, I would have come here and made camp; a place like this is bound to be crawling with game. It's likely that the tribesmen who once lived here knew it, but we must be the first white people ever to come here."

Beyond the escarpments, the Luwegu crosses a flat plain, but in the southern distance can be seen the sudden mountains where the river descends from a steep-sided gorge. The Wandewewe Hills beyond are part of the Luwegu watershed, but most of its tributaries descend from the Matengo Highlands west of Songea. These elevations on the great inland plateau are drained by the long rivers that flow northward, following the tilt of the plateau toward the Rufiji. The Luwegu and the Mbarangandu run almost parallel, and both are characterized by broad canyon valleys between steep cliffs of reddish sandstone that rise on both sides to long flat-topped plateaus.

We follow a steep buffalo trail down the escarpment, crossing the plain of baobab and pushing more buffalo out of the dense shade of a karonga. Then we climb again, still too far to the east, working our way around more hidden buffalo; in the very steep, hot, thorny going the Ngindo are exhausted, but on account of the buffalo their nerves give them the energy to keep up. Over the centuries the elephants have found all the paths of least resistance, and Goa follows their clear trails up and down the broken country, but by midday two of the porters are "pretty well knocked-up", as Brian puts it. Because there is no shade where we strike the river, we continue along the hot slow sand, enter the bush again and find ourselves between an agitated hippo and the river.

For all one reads about lion and elephant, buffalo and rhinoceros, it is probably this vast water pig of other ages, with its immense jaws and long shearing teeth and unexpected speed, that kills more people than any other animal in Africa, and it does so most often when, as now, it finds itself cut off from sanctuary in the water. We stand a moment, at a loss, listening to this beast's companions disporting in the river, and then Goa leads in a polite circling maneuver, working inland to give the hippo a clear path. Misinterpreting this move, the hippo retreats inland a little, too, then stops to glare again, coming fully about to face us as it does so. Because we are all hot and tired, and no solution is in sight, we cross quickly between the hippo and the river as Goa calls to the young porters to keep up.

In the dense river thickets once again, the porters keep up smartly without being told. Brian kicks at fresh manure and mutters, "Liable to turn up a buffalo just now . . ." He has scarcely spoken when the porters yell and scatter, dropping their loads. Goa whirls and raises the rifle; we had passed the buffalo, which then burst from a thicket by the river only a few yards from the line of men. It does not charge but runs off in the direction we have come, as all the porters squeal with nervous laughter.

Never having been on safari before the young porters are inexperienced with big animals, but this has not always been true of the Ngindo. Though they practice a subsistence agriculture – growing mostly maize, millet, and cassava – and keep a few goats and chickens, the Ngindo were essentially hunter-gatherers until this century and to this day have retained the small, slight stature of bush hunters. Probably they made small impact on the wildlife populations, since apparently they were always few in number, living in small, semi-nomadic groups along the rivers. A Portuguese document of 1616 mentions the sparsity of the inland population in this region, referring to it as a *"terra deserta"*, and human numbers were maintained at a low level by the constant attrition of the Arab slavers, the Ngoni Zulu, and the Germans. Though some Ngindo must have been taken as porters and slaves, they offered no resistance to the caravans of the eighteenth and nineteenth centuries that passed through on the way to the interior from Kilwa; where chance offered, in fact, they stole one another's children and bartered them for salt, cloth, and the primitive firearms that still serve in out-of-the-way places as items of prestige, and thus made at least a minor contribution to the desolation and disruption of the countryside, the pillage and burning and inter-tribal raiding, and the destruction of wildlife that the slavers and ivory-traders left in their wake almost everywhere throughout the country.

Though local tradition says that the Ngindo people were always forest hunters, it may be that this was a consequence of chronic disruption and flight. "We do not stay long enough to eat our own mangoes", is an Ngindo proverb. The original Ngindo homeland is thought to have been further south, extending across the Ruvumu River into what is now Mozambique; apparently they were pushed northward by the Ngoni Zulu who swept up from southern Africa in the nineteenth century. Eventually the fierce Ngoni, with their well-organized militia, came into bitter conflict with the Germans who were settling this region from the north, and the hapless Ngindo were caught in between. Having gone to the Germans for protection against the Ngoni, they found that these white men were still worse. *Wanatoka wafako, wanakwenda waziwako*, they said, in mourning for themselves: They go from where there is death to where there is burial.[2] The Germans needed slave labor to make cash crops out of their cotton, and malingerers were given twenty lashes. These beatings inspired such resentment that finally, in 1905, the Ngindo people at Kabata, not far east of the Selous Reserve boundary, rebelled against an order to pick cotton. The revolt spread to the Lung'onya River region, where a witch doctor produced a magic water to protect Africans from European bullets, and soon five Germans, including a bishop and a nun, were destroyed by Ngindo at a place called

Mukukuyumbu, near Liwale; others were killed by Pogoro people near Madaba, and it was at Madaba that there commenced, in November of that year, the efficient and thorough suppression of what came to be known as the Maji-Maji Rebellion. The Germans are said to have made a point of executing the eldest son of every family in the region, and at least 100,000 Africans were slaughtered. (In this same period, the Germans were carrying out repressions of a comparable brutality among the Herero of what is now Namibia.) Everywhere villages and crops were put to the torch, with no respite for the planting of new gardens, and in the next three years as many thousands – hundreds of thousands, some authorities have said – died of starvation, the humid skies were dark with vultures and a shroud of fire, and lions prospered. In 1908, an old man named Sulila at Masasi made a song recounting the dark years of his people:

> Then comes the war of the Mazitu [Ngoni]; guns are fired by the Germans; then they ran away. But the Germans came; it was dangerous to see. The bush was burnt; the goats were burnt; the fowls were burnt – the people were finished altogether. The tax came up . . . still they were not satisfied. Mr. Sulila telegraphed to the District Commissioner, "He may skin me to make a bag for his money. Now I am tired."[3]

An historian wrote, in a paraphrase of Tacitus, "The Germans in East Africa made solitude, and called it peace."[4]

Just after five, we make camp beneath a big tamarind at the river's edge, opposite a grove of high borassus. Upstream, a solitary bull elephant wanders the bank; further on an eland emerges from the trees, very pale against the dark greens of the river forest, and a cow elephant and calf stand at rest in the late afternoon sun, as if lost in some long twilight meditation. A harsh racketing downriver is made by a pair of huge Goliath herons, perfoming a courtship ritual in the shallow water: it seems appropriate that the largest members of their worldwide families, the Goliath heron and the giant kingfisher, are still abundant on these big wild rivers, though no longer common elsewhere in East Africa. The herons display to each other with broad wings and nervous dancing as a third heron, to all appearances an injured party, turns its back on them and sits in a hunched position on a sand bar, staring toward the north.

Kazungu brings us mushroom soup, and Brian gives his bowl to me; he cannot eat it. In 1959, he says, he came down with serious stomach pains, which he did nothing about for four years. "Had knotted guts whenever I ate anything; couldn't eat fresh fruit or vegetables at all. Finally got down to about one hundred pounds." Not until it seemed that he might starve to death was he persuaded to go to the tropical disease hospital in London, where his ailment was eventually diagnosed as *Histoplasma Duboisiei*, the first recorded case in Africa of a rare gut fungus that is not uncommon in the Persian Gulf. "All they had were some experimental drugs, which they did not hesitate to recommend since otherwise I would be dead within six months. And the drugs worked, because here I am, but I really couldn't eat properly until 1973, when a friend of ours up in Nairobi recommended yogurt. Cleared me right up in two weeks – the only thing I can't eat now is mushrooms."

In fact Brian still eats very little, and seems to prefer the gray oatmeal gruel and rice that tided him over his long illness. Not being the sort to concern himself about other people's preferences, he discarded almost everything with taste in it that Karen Ross had set aside for our safari, saying that we must travel light. ("I tried to slip in a few goodies for you," Karen warned me, "but he tossed them right out again.") Brian sees no reason why I should not adjust my habits to having two meals a day. "People worry too much about food," he says. "Afraid of going without, so they eat too much, women especially – their main occupation, I suppose. Especially here in the tropics, where habits are apt to be sedentary, it is better to eat less." And so the single bowl of gray *porrigi* at mid-morning represents the midday meal as well as breakfast, and the rice with beans at supper is the one full meal of the day. I don't bother to protest that his bloody tent weighs a lot more than the discarded food: two meals suit me in this heat, and eaten outside under the trees and stars, the simple fare tastes very good indeed.

I ask Brian what the tropical disease hospital said about the yogurt cure for *Histoplasma Duboisiei*, and he says that he did not bother to report it. Taken aback, I ask – not entirely insincerely – if he does not wish to help his fellow man. "Not really," he said, not entirely insincere himself, and again I hear that echo of Ionides: "The convenience or profit of others is a matter of supreme indifference to me unless it happens to coincide with my own."

In our camp tonight, Kazungu writes:

In the morning I made tea while other people packed up the camp – a tent for the two white people and a ground sheet for the rest of us. Then we began our safari. As usual we take our breakfast after two hours. We came across six elephant with calf and had to pass far away; as usual, an animal with baby is very fierce. We saw four buffalo, which are bad animals; he doesn't care about anything or anybody, but does what he likes, you cannot trust him. I fear the buffalo more than anyone . . .

We made a big fire because of wild animals like elephants and hippo so they would not come near. Most of riverside is path of wild animals to come and drink water. I didn't get good sleep because I have never slept outside without a tent.

It is always Goa who puts up the tent, assisted by one of the young Ngindo, and it bothers me slightly every time they do it, probably because the Africans do not have one. It is true, of course, that they could erect the canvas kitchen fly brought for that purpose, but they prefer to use this as a ground cover, sleeping in a tight row on top of it and keeping the fire going all night long. They are uneasy about elephants as well as lions, but elephants rarely wander into camp except in parks. As Brian says, "The elephant is very considerate, really. Rhinos blunder in sometimes, and lions come about, but a round from the shotgun usually sees them off. Now a man-eater, of course . . ."

In the distance rises the early morning sound of the ground hornbill, the remote dim hooting of a woodland spirit – *poo-too, po-to* – a lugubrious ghostly reverberation that seems to emanate from a cavern under the earth. In West Africa this bird is considered sacred, and Kazungu and Goa say that if you kill one, the young ones will appear by your house and make this sound – *poo-too, po-to* – and very soon you will take sick and die. The Ngindo say that Kazungu and Goa are mistaken: if you should be foolish enough to kill the *nditi*, as the bird is called in this part of Tanzania, you will die without further ado.

We make good time over the river plains – we are now well south of the black rock escarpments – passing large tame herds of wildebeest and impala, slowing only to make our way around the many elephants, not less than six groups in the first few miles, although there are never more than five together. None seem bothered by the file of men, there is no threat display whatever, and most of them do not move away as we go past. Buffalo sign is copious, but the buffalo remain hidden, as do the rhinoceros and the lion: last night I heard a lion roar toward dawn – I have heard lion almost every night since coming to the Selous – but in this long-grass country they remain unseen.

A wood of "silver trees" *(Terminalia sericea)*, as Selous called them in his journals, descends to broad pans of black cotton clay twisted up into hard ruts and potholes by huge megafaunal feet – a quagmire during the rains, now petrified into near-stone that is very hard on the bare feet of the porters. Then we are in river forest once again, and here Goa slows a little, feeling his way into the gloom with hand held out in front like a moth's antennae, leading the file in a winding course to avoid a confrontation with the depositors of all this fresh and shining dung. In pockets of sun, I find myself alert to the white butterflies, cat musk, thick blood-red kigelia blossoms with long yellow stamens, and the somber hum of bees.

In mid-morning we come to an area that has been burned over. The fire is recent; there are no sprouts of green tussock, only white remnants

(Right) Departure on the foot safari.

[*144*]

(Over) Little bee-eaters.

on the blackened ground – the bone-white shells of millipedes and giant land snails unable to escape the sweep of fire. Antelope droppings are baked a strange bluish gray. Goa locates a simple cooking fire in the thicket on a mound, and we find the place, scraped bare of stones, where men have slept – the first and last sign of man's presence that we would come across in this country south of Mkangira.

After a silence Brian says, "Looks like they were here about a week ago." Perhaps he, too, is thinking that the animals seem very tame for a country in which poachers have been operating, for now he says, "It's possible, of course, that it's *tambika* – that's ancestor worship, from the days when people lived along these rivers. They come out here sometimes to *tambika*, or at least they did. On the other hand, poachers will use that as an excuse: 'We're just a few ancestor-worshippers who stumbled upon all this ivory we're carrying.'"

Over mid-morning *porrigi* and tea, Brian talks about the footpath that crossed the Selous Reserve between Liwale and Mahenge, beyond the western boundary. The Ngindo of Liwale have relatives at Mahenge who were cut off when Ionides's expanding elephant reserve was joined to the original Selous, and to visit them without trespassing in the Reserve meant a journey on foot and by jitney of at least eight hundred miles. Since even those who could afford the time could not afford so many jitney buses, it was inevitable that people would cross illegally, and rather than have them scattered through the bush and perhaps remaining there, Ionides had made it legal to use this footpath, which was about 120 miles in length. In Brian's opinion, this path a few miles to the south was probably the access route for whoever had made that camp fire.

"When they come in this far," he said, "they're after ivory and rhino horn; they can get meat much closer to home, most of it legally, because the country here in southern Tanzania is still wild. This isn't Kenya. In most of Kenya, until recently at least, almost the only game that people could find in most areas outside the parks was a few small antelope. Of course the poachers will shoot game, too, to live on, the way we're doing ourselves on this safari, but that's not why they're here."

The Ngindo hunters are leaders in their villages, as is the case among Goa's people; they are "poachers" only to the white men, whose demand for rhino horn and ivory is what has put the animals in danger. "Often a village will have one great hunter," Brian relates. "The people chip in and buy him a rifle and arrange for his bearers and assistants; others may also buy a gun and have him do their legal killing for them. He's not somebody out of work, or outside the law in any way; invariably, he's a respected member of the community, with wives and children and shambas and all sorts of prestige, which he deserves, because he takes risks and he knows his business. That's why it's hard to get people to testify against him. When I sent a man like that to jail, I used to go and pick him out again in about six months and have him come work for me;

he would always be capable, and would know his region of the bush in great detail.

"Of course, people coming in this far have the rivers to contend with. In the dry season, they can wade across if they want to risk the crocodiles, but in the old days, at least when there was patrolling, they came mostly in the rains, when they knew that the tracks would all be mired and our machines unable to move. In a river as large as this, they'd use a *kungwa*, which is essentially a section of muyombo bark about nine feet long, the half-round of a big tree or all of a small one split along one side, with ends plugged up with bark and mud. Pretty rickety affair, but it usually got them over. Up on the Kilombero, of course, the poachers coming in from outside would come down in canoes, especially when the crocodile trade was in full swing." Brian grinned a little to himself, remembering. "Once we were trekking up the Kilombero – we'd heard there was poaching up that way – and we saw that the poachers had their camp out on an island, and we had no boats. So we sent one of the game scouts up the river, pretending to be a poacher himself, and yelling across that a hippo had overturned one of his canoes and a man had drowned, and to please send help. When the canoes came from the island, we arrested the rescuers, then used the canoes to go back to the island and arrest the others, though not without a fight."

Until 1958, when Alan Rees became warden of the western Selous, the poaching there had been almost as rampant as it was in the north, near the towns and the main road; in fact, the Reserve was overrun by meat-hunters, often outfitted by local traders, and the slaughter of game was so widespread that certain regions have not fully recovered to this day. Not until 1961, when the Selous Reserve administration was coordinated under a new chief warden, Major B.G. Kinlock, was poaching brought under control in the north and west: between mid-1962 and the end of 1963 some 1600 people were convicted for offenses in the Game Reserve; many tons of wire snares and miles of fencing were seized and destroyed, and more than 200 firearms were confiscated, together with an enormous collection of spears and bows and arrows. From that time until Nicholson's own departure in 1973, organized poaching was no longer a problem in the Selous.

On the river beach at the next bend, where two hippos stand facing each other as if made of stone, a line of *siafu*, or soldier ants, is moving along a narrow tunnel just under the surface of the sand; the tunnel is roofed with a sand crust, but here and there the crust has fallen, exposing the glistening nerve of ants as they hasten away into the thicket.

"I remember the old days on the farm," Brian said. "We'd have to camp out when the ants came, because there were no insecticides or anything. But as soon as everything was eaten, they would go away, having cleaned the place out down to the last cockroach; wouldn't see a rat or a mouse for the next six months. There's nowhere to hide from them – swarm over everything." Recalling how back at Madaba the *siafu* had driven Tom Arnold from his tent one night, Brian grinned. The only thing that could stop these *siafu* was another ant, the *sangara*, a very swift yellow-brown species which he had pointed out at Kingupira; they appeared to run nimbly over the backs of the slower *siafu* and spray the line with some sort of formic acid.

On the first day of our journey Brian had mentioned that his red sneakers were giving his toes hell and cramping his feet too, but he ignores my suggestion that he put on socks or plasters – "No room", he says – and he has no other boots or shoes to take their place. I have no spares either; in the cause of traveling light, we are sharing one small duffle that contains everything we have brought, including flashlights, Brian's cigarettes, and my notebooks. Yesterday Brian was obliged to bandage his little toes, and this morning, after the first two hours, he borrowed my knife and cut holes in his sneakers to keep them from doing him any further damage. As it is, we shall have to quit in the early afternoon, and tomorrow – a day early – we shall leave the Luwegu and head east toward the Mbarangandu, before the distance between rivers grows any wider. This isn't because of Brian's feet, or because we have not seen animals – the plains game were all there this morning as well as a number of elephant and buffalo – but because of that burning, which we have assumed was done by poachers. With the sight of that fire-blackened land, of that transgression, a sense of the vast silent Africa, "the Old Africa", was dissipated, and the Mbarangandu will be our last chance to restore it.

This afternoon we take shelter from the sun in a bush orange grove by the Luwegu, drinking our tea and chewing on the long hard strips of buffalo biltong. Eventually the biltong proves too tough for the false front tooth I had installed eighteen years ago to replace one broken on an expedition in New Guinea. The damned tooth shatters to pieces in my mouth, and I spit it out in consternation; as in New Guinea, I am pretty far from help if the thing acts up. Brian is not the least bit sympathetic; in fact he laughs. "Makes you look tough," he observes, "like you really mean business." And he removes his own false teeth, leaving gaps on both sides of his incisors that give him the aspect, as he says himself, of a huge rodent. Turning his back, he suddenly whirls and stares at me over his shoulder, letting the two rodent teeth emerge on his lower lip. "This is your captain," he says, and bursts out laughing, claiming that he does this sometimes with strange passengers in his airplane. Though I doubt this story, it doesn't matter, since he has me laughing, too. "If you hadn't been going at the bloody biltong like a hyena, instead of chewing it off in delicate bits like me, it would never have happened," Brian says, and we laugh anew.

X

A light unseasonal rain that fell last night is attributed by the Warden to the bush fires that in these months are set all across Africa in the *miombo* belt, from southern Tanzania and northern Mozambique to the coastal forests of Zaire and Angola. The morning is heavy and humid after the rain, and two of the porters, complaining of headaches, are given aspirin. Behind their round dark aching heads, a flight of egrets passes down the early river.

During the night, a small leopard made off with one of Kazungu's sneakers, which on this foot safari have replaced the green Wellingtons he was wearing for a while as a precaution against cobras. As the only staff member with shoes (although Goa uses rubber sandals that make a faint snick-snick as he goes along) Kazungu is understandably upset, but the cat prints are clear in the damp sand, and he soon locates the spat-out sneaker a short distance back in the thickets.

This morning we abandon the Luwegu in order to cross the ridges toward the east and explore an unnamed sand river that comes down off a high plateau. On the north end of that plateau, high above the dry savanna woodlands all around, is the large pan that the Warden remembers as "crawling with elephant" and other animals. He is eager to revisit the nameless plateau and its pan, which he discovered in his last years in the Selous.

An hour is needed to cross the flat river plain of the Luwegu, on the east side of which a pair of Bohor reedbuck start up from high grass along the wood. The going is mostly very fast, but because of large animals we travel no more than a few miles in an hour. Buffalo tracks and buffalo manure are copious, and so are the attendant flies that do not sting but alight damply on the eyes and mouth. Then we are in thickets and karongas once again, and Goa, squatting to see beneath the bushes, craning, listening, picking out the big and silent shapes that watch us pass, must move circuitously and with more care. Brian, too, is wary and alert, kicking apart a fresh elephant mound, then stooping without breaking stride to judge the proximity of its maker by the degree of warmth that rises to his fingers.

In a sand river – and he thinks this is the one that has no name – we meet a cow elephant with half-grown calf. The cow goes off into the thicket to our left and the calf dawdles, then blunders toward the right, starting to bawl. Unaccountably, Brian and Goa move between the animals, and a moment later the cow, no longer visible, is bellowing from a short distance away; reverberations from the thicket make it clear that other agitated pachyderms are behind her. Goa turns quickly, taking the shotgun and giving Brian his own Game Department rifle. ("He knows I am more skillful with it," Brian explained later, "so he doesn't mind.") The porters rush forward in a covey to take up positions behind the guns. After a mild demonstration charge that brings the guns up, the cow wheels and goes trumpeting off with the others; probably it was not that dawdler that had concerned her but a much younger one too low down in the bush for us to see.

A mile away, vultures spiral upward from a kill, and we have gone less than a few hundred yards when Goa's deep and urgent voice says, "*Simba!*" Two lionesses, then two more, shoot out from beneath an enormous tree fallen into the river bed perhaps thirty yards away; they are followed by a big growling male with a fine mane which accelerates as it sees the file of men, its big paws scattering hot sand. The lion bounds across the river bed and up the bank. The whole bend of the river stinks of lion, and there are print patterns of the litter of small cubs, which must be lying hidden just close by. Brian says, "I seriously doubt if those lions have ever seen a man before; even the poachers stick to the main rivers. Yet look at how fast they shot away! What makes them run off like that? Quite interesting, really. If we'd come up on them in a Land Rover, chances are that these wild lion would be just as tame as those lion in the parks. It's the cars that fool them; they can't seem to identify men in cars. If you're up in a tree, a lion will recognize you straight away, even though that's not the way he's accustomed to seeing you."

A lone African hare crosses our path in its age-old silence, and in the thicket, banded mongoose skirl and squabble, in furor over some edible find. Every little while we pass the dung-spattered double scar on the dry ground that marks a rhino scrape; although we have seen many fresh scrapes in the past few days, the great primordial beast itself has remained hidden. Soon the porters take their rest in a grove of the sand river, which twists and turns back and forth across our course toward the south, and Goa and Kazungu dig down in the dry sand until the hole fills with the clear water that will be used for morning tea and porridge.

In the early afternoon, blue water glints in the eastern distance; at this time of year, when all but the main rivers have gone dry, this can only be the Mbarangandu, which we had not expected to see until tomorrow. If the two rivers are so close – not much more than twelve miles – then we are further north than we imagined, which means that we have struck almost due east, instead of east by south as we intended; we have no compass, and throughout the morning the sun that would have given us a bearing has been hidden in the fire-shrouded sky. We turn due south. I am glad of the miscalculation, which brought us to this place at the right moment: we have hardly seen the glint of river when Goa points out two rhinoceros, a mile away down a long slope of the savanna; one fades quickly into the high grass but the other lingers a few moments, turning broadside, before barging off into the bush.

I yip with pleasure, and Goa is delighted. To Brian he says, "The only thing we must show him is leopard!" Goa has a sudden fine full smile that sends wrinkles back on his tight hide across the high cheekbones to the small tight ears: the old hunter looks like a hominid designed for passing through bush quickly and quietly, catching nothing on the thorns. Never watching the ground for vines and holes and sharp grass-hidden stones but seeming to drift over the earth, he scans the terrain with those yellowed eyes that see so much, on all sides and far away.

Old elephant paths of other seasons cross the rolling hills of high bronze grass, but there is no sign of animals whatever except for two pretty klipspringer, tawny and gray, which prance along the black granite rim of a low escarpment; from this black rock, in the white sun, so very hot that it bakes our feet, a dark thing flutters up like a great moth – the freckled nightjar, which makes its home on the black outcroppings of stone.

Here and there on these black platforms, emerging like low domes out of the grass, lie shards of quartz, agate, chert and other stones, apparently brought here to be worked by stone tool cultures, for many are distinctly flaked, with the characteristic hump that betrays the method; occasionally I pause long enough to stuff one or two of these ancient tools into my pockets, including round tortoise cores flaked all the way around the edge, like those Hugo and I had found on the ridges north of Mkangira. The heat of the tools, and the feel of their great age, is somehow satisfying and profoundly reassuring, as if we had passed into another age, as if those Stone Age men had paused here yesterday, taking the sun god's name in vain as they cursed their sharp, obdurate stones, scowling in the heat.

Cutting across a series of wide bends, still heading south, we come down to the river. Two waterbuck, two wart hog, and two buffalo wait on the bar, like creatures left behind by Noah's Ark; the buffalo refuse to give ground at our approach, although we shout. We have circled wide and now stand in the shallow water, trying to ease them up into the thicket so that we may proceed along the sand. "Don't want to make them think they're trapped," says Brian. We can go no closer to the lowering brutes, which have backed up with their spattered rumps against the bank. Before the wrong move can be made, the buffalo wheel suddenly and plunge off up the bank into the thicket, agitating a group of elephants that we had not seen.

In the shade of a big butterfly-leafed piliostigma we stretch out on the cool sand, and I listen to the young Africans behind me; they are still excited by our encounters with the *tembo* and the *simba,* and describe to each other in dramatic tones and with nervous squeals how Mzee Goa and the Bwana Mkubwa (for they are too young to have known Brian as "Bwana Niki" and refer to him as the Head Bwana, the "Big Bwana") raised their *bunduki* to protect our lives.

While Kazungu busies himself over his pots, Goa walks inland and sets fire to the bush, then wades across the shallow Mbarangandu and fires the high grass on the far bank; in a few minutes, the clear African day beside the river is despoiled by a crackling roar and columns of black smoke, which stings the eyes, and stinks, and dirties the sky. A pair of hawks are circling a tree where the accumulated grass and deadwood is feeding fire that booms and reverberates as it moves away; perhaps their nestlings have been singed of their feathers, and even now nod just a little, blackening in death.

Brian, sensing my disgust, insists once more that early-season burning is essential to wildlife management in the Selous. There are moderate rains in the months of December, less in January; the heavy rain that renews the vegetation is concentrated between late February and early May, when twenty-five to thirty inches may fall here in the south and twice that amount in the Kilombero region and the west. The two rainy seasons tend to merge into one long one, in which the grass grows very high, coarse, and unpalatable, and smothers the wild pasture for the remainder of the year; the use of fire as much as the removal of the inhabitants is responsible for the fact that the game population of the Selous is many times greater than it was when the first Europeans came into this country.

No doubt Brian is right in terms of game management; it is the necessity of all this "management" that I resist. One day I ask him if any of these fires ever occur accidentally, through lightning or other random events, and he says that he thinks it very unlikely; even if they did, they would not travel far. This poses a question that neither of us can answer: since it is thought that this "dry forest" is a recent habitat type, created, perhaps, by the fires of those early hunters who left their flaked tools on almost every high place and granite outcrop in this landscape, and spread and maintained by human activity ever since, how is it that such creatures as Lichtenstein's hartebeest are endemic to the *miombo*, since they must have evolved many thousands of years before mankind had fire at all?

As we talk, Brian makes an odd drumming with his fingers on a log, two soft beats followed by two hard, and noticing that I notice, he says *"Mgalumtwe.* That's the local dialect for 'A man has been eaten.' They drum this message on a hollowed-out log with a bit of hide stretched over one end: M-ga-LUM-TWE! The villagers come to the call of the drums, armed with spears and bows and arrows. Sometimes they harass the lion but they rarely kill it; it just goes off, more dangerous than before. That man-eater might be raiding in an area of a hundred square miles, which makes it very difficult to come up with, especially when you're traveling on foot. It might take two people in two nights in the same place, then disappear entirely for two weeks, presumably taking animals instead. Often man-eating is seasonal, during the rains; in the dry season, when animals are concentrated near the water points, these lion seem to prefer animals, returning to human beings in the rains when the animals scatter. But man-eating was certainly most prevalent where game was scarce because of human development; the lion that were unable to catch what game was left – especially lion that were old or crippled – turned to human beings. I remember one I shot, still on the man whom it had killed outside his hut and dragged into the bamboo. That lion was in terrible condition, half-starved really, due to porcupine quills and an infection in its throat that kept it from swallowing. It could only take little bits at a time, which was why it was still feeding on that man when I arrived.

"But most man-eaters I saw were in good condition, and they were wary. Baits rarely worked on them, they were too clever. You had to wait until the next person was taken, doing your best to persuade the villagers not to drive it off the kill but to let it gorge itself. After that, it wouldn't go too far before it fell into a heavy sleep, and I'd have a chance to reach the place before it woke up again and moved away. Tracking it, you'd find places it had lain down and then got up again, until it found the place where things felt right. Usually that was in deep thicket, and sometimes all you could see when you crept up was a patch of hide. Finally you had to shoot at that and hope the bullet would disable it. Otherwise, you might have it right on top of you."

One problem was the superstition among villagers that a man-eater must be a *mtu-ana-geuka-simba*, literally, "a man-turned-into-a-lion", a witch with whom it was dreadfully dangerous to interfere. This superstition was often shared by the local game scout stationed in that village or sent out to dispatch the lion – including the mighty Nonga Pelekamoyo, or Take Your Heart, uncle of old Saidi (and also uncle of Rashidi Kawawa, Tanzania's first Prime Minister after Independence, and currently Minister of Defense). This two-hundred-pound Ngoni – the one who had set fire to all the Ngindo settlements on Brian's first tour through the region of Ngarambe – was sleeping in a bamboo stockade in a village beset by a man-eater when the lion burst straight through the bamboo in order to get at him. Miraculously the lion seized, not Nonga, but the wooden bed that the villagers had provided for their savior. The lion actually dragged the bed clean out of the stockade, and Nonga Pelekamoya, managing to roll off, escaped unharmed. But afterward Nonga refused to consider any further dealings with this *mtu-ana-geuka-simba*, and turned the whole case over to Bwana Nyama.

Brian reckons he has killed about fifty lions, of which perhaps nineteen or twenty were man-eaters; the rest were stock-raiders, which usually got that habit from feeding on dead cattle after drought or plague. Asked if fear had ever been a problem, he thinks a moment, as if such an idea had never occurred to him before. Frowning, he says, "One is bound to be tensed up, of course, but if it was *really* fear, you wouldn't bloody well do it, especially after you've put a bullet in some dangerous creature only to have him go thrashing off into the thicket. Then you've got to start all over again, and it's a lot worse than before." He shook his head, and changed the subject. "I recall one lioness over here on the Mbarangandu that jumped out and scattered the porters, then came towards me. I knew she was not a man-eater, in a place so remote from human habitation; I thought she must have cubs, so I didn't shoot, just held the rifle on her, backing up slowly. She kept on coming, keeping the same fifteen yards between us, snarling and thrashing her tail. And then, when she figured that her cubs were safe, she turned suddenly and bolted for the long grass. Quite interesting, really."

"In the north," Brian says, "man-eaters are rarely a problem, but here in the south they still occur regularly. Not so many in the settled areas any more, because the lions themselves are dying out, and there is still enough game around to feed those that are left. But two or three people have been taken along that foot path" – and he pointed south – "between Liwale and Mahenge. Except for that porter at Kichwa Cha Pembe, I've never lost any of my people to a lion, but you have to be careful."

By nightfall the humidity has lifted, and the flying clouds part on a cold full moon. All around the horizon, as the wind chases them, the flames of Goa's fires leap and fall in the black tracery of trees like a demonic breathing from within the earth.

Lying out under the stars, we reminisce about friends we have in common. When first approached about joining this safari, I had been told that Brian Nicholson had invited along Myles Turner, now a game warden in Malawi, who had befriended me in the Serengeti in 1969 and 1970; this evening I expressed regret that Myles was unable to join us after all. Brian stares at me. "I thought it was *you* who had invited

Myles!" he said. I shake my head: I had been told that Myles was an old friend of Brian, and the suggestion was made that I could get a lot of good material sitting around the campfire at night listening to the two wardens talk about the late great days when the native knew his place. Although Brian knows that I am teasing him, he grins.

"Well, I've known Myles for a long time, that's true; I've known him since about 1948. Good hunter, too – very patient and painstaking. And as I recall, the first gun I possessed, some sort of air gun, once belonged to Myles. But I can't imagine what Myles and I would have reminisced about; never did one thing together that I can recall. Last year Myles was in Nairobi, and I asked him to stop by the house, talk over this safari. He said he was very busy, had to go to Nyeri and so forth, and as it turned out, he had no time for me when he got back. Can't say I was surprised when the news came that he couldn't get leave after all. In the old days, when he was a warden in the Serengeti, he used to say, 'I'd give anything to get down there and see that bit of Africa of yours.' I invited him, and more than once, but he never came."

Another friend of Brian in the early days was a young Provincial Forest officer named John Blower, a wanderer whom I crossed paths with many years later, in Ethiopia, and again, years after that, in Nepal. Blower had been fascinated by the possibilities of the Selous and thought that the Lung'onyo River region might be best protected by making it part of the Forest Reserve. At one point, he made what Brian calls "a hell of a trek" from the Kingupira region west to the Ulanga River, then south to Shuguli Falls, then up the Luwegu River to Mkangira, and from there back to Liwale. "Half-killed his porters. He'd set out in the morning and never look back, and some of these chaps were still turning up weeks later. Couldn't hire a porter around the Liwale area for six months afterward." Hearing of this, Ionides had been furious that a walk through the Selous should have been made without his permission, and subsequently, at some sort of Game Department function in Arusha, he asked a young man if he had ever come across "this bastard Blower". It was Blower himself, of course, whom he addressed, and after a decent interval, they became great friends.

In 1953 Nicholson was summoned to Kenya to serve with the Kenya Police Reserve, patrolling against Mau Mau terrorists around Nanyuki. By 1954, when he was called a second time, the hard-core Mau Mau had retreated up into the Aberdares, and Brian served with the Kenya Regiment in field intelligence operations about twenty miles northwest of Nairobi, where he participated in the "pseudo-gang" operations. "We'd black our faces and go out at night, wearing old clothes, and an African whom the Mau Mau did not know had turned against them would lead us straight into their camp. He'd do the talking to get us past the guard, and once we got into the middle of them, we'd shoot the place to pieces. That's what really won the war; it completely confused them, they never knew who was for them and who wasn't, and there were cases toward the end in which two genuine Mau Mau gangs were shooting up each other. John Blower was sent up there on the same thing, though I didn't see him. I was with Billy Woodley and several others; we were very close friends in school, Billy and me, and we still are." Remembering something, Brian smiles. "When we were kids, we used to go hunting out on the Aathi Plain, where Embakasi Airport is today. One day we were trying to stalk some tommies, using cattle to cover our approach – no luck at all. Then more cattle came along led by a prize white heifer, and damned if I didn't forget that gun was loaded. I said to Billy, If that was a buffalo, this is how I'd shoot it – BLAM! Down it went! That must have been thirty-five years ago!" As Brian says this, he sits straight up, more startled by the passage of time than by the trouble he had brought upon himself with that fatal shot.

At dawn, the smoke of Goa's fires has gone, the air is clean and cool, with the wind from the southeast; the prevailing weathers all across East Africa derive from these easterly trades off the Indian Ocean. Until today, it has been hot an hour after sunrise, reaching 100 °F, or so we estimate, late in the morning and maintaining that temperature until mid-afternoon. For those who must carry loads through thorn and tall dry grass, over black granite and the lava-like black cotton of the *mbugas*, it is just as well that on most days we travel for no more than five hours. Since the porters are well rested every afternoon, their spirits are high – so high, in fact, that occasionally it is necessary to damp them down. "*Usi piga kilele!*" Goa whispers at them, turning on the trail. And they do not mimic him or mutter, only smile a little, walking along under the awkward loads with the swaying elegance of the women in their villages, arms close to their sides but hands curved out, fingers extended, the loads clinging somehow to their heads. Only after moments with big animals do they hoot and chatter, letting off steam, and if this makes me smile, and I cannot hide it, they then have the excuse they need to squeal with laughter.

Although Brian is fond of saying that these porters can't compare with those on his old staff, who were "trained up to it", they win his grudging respect as the days go by. "They're a very good lot," he acknowledges. "Out five days, and not a single complaint yet." Even the saucy and single-minded Mata, from whom at least impertinence had been expected, has decided to comport himself as a professional porter on the basis of his one previous safari, and sets a stern example for the others. (Only at Mkangira, in the *ngoma* put on by the staff, did Mata display his cynical opinion of the hospitality, greetings, and thanks to the white visitors that the songs were intended to convey. In woman's costume, Goa's straw hat pulled down rakishly over one eye, he went swanking in and out of the line of porters as they danced and chanted, mimicking their thanks to the White Bwanas with squirming hips and raised prayerful hands and eye rollings of burlesqued gratitude – *Asante sa-a-na!* – in a parody so wild and deadly in its execution that whites and blacks alike giggled uneasily, not knowing where to turn. Mata himself had laughed openly into our faces, and his revenge was the artistic high point of the *ngoma*, surpassing even the strange dance of Abdallah, who kept time to the tom-tom while hopping on his head and elbows all the way around the fire circle.)

Leaving behind the smoldering black land of Goa's fires, we wade the river and head south through long, rolling savanna hills between the Mbarangandu and Njenji rivers. The morning remains cool and pleasant, with no trace of humidity, and the high grass, bronze and shining, flows in the south wind and early light. A flight of the large trumpeter hornbills lilts along the dark lines of big trees where a karonga descends toward the river, and a solitary white egret stands immobile in the rank green margins of a spring; this is the common egret, a cosmopolitan species, the only bird in the Selous that is also common in North America.

Coming up swiftly over a rise, we run into an old bull buffalo perhaps thirty yards away. As Brian seizes Goa's rifle, the bull rears his snout, seeming to glare at us past flared wet nostrils, big horns shining; the depthless black eyes never blink in the long moment that it takes for nose and ears and eyes and modest brain to weigh the choice between attack or flight. We hold quite still. Given enough time and space, the buffalo will almost always take the most prudent course, and after a few seconds of suspense this one, too, gives the heave and snort that accompany these bovine decisions and goes crashing off downhill into the thicket.

Apart from this buffalo and a few small groups of elephant, animals in this long-grass country have been scarce, as if all the waning energy in the coppered hills and yellowed trees and sinking rivers of September had been distilled in the fierce greens of the parrots and the paradisal blues of the brilliant rollers. But on the far side of a rise, in an open hollow, the missing animals have taken shelter from the silent landscape – impala, wildebeest, wart hog, zebra, and elephant, with a flock of the huge *batutu* in attendance. Led by the hornbills, all but the elephant stream away uphill toward the blue sky, striped horses shining in the high bronze grass.

Goa follows the old paths of the elephants, which follow the ridge lines and avoid the stony depths and thorn of the karongas; it is pleasant to sense that most of these neat trails, two feet across, from which all grass has been worn away, have never been seen or walked by human beings. Eventually a path descends toward the river, and from a bluff we contemplate an elephant with calf, at rest on a clear shoal in the middle of the glittering Mbarangandu; in the new light of the river morning, they seem to dream, lulled to forgetfulness of where they might be going by the clear torrent from the southern mountains that casts sparkling reflections up the gray columns of their legs. We wait. Others have crossed ahead of them, feeding calmly in rank canebrake by the river, and when the cow and calf move on, we wade out into the sun-filled flow and continue southward.

Before long we are stalled by another elephant, this one a young bull standing in the grass of the river margin. Since more elephants are visible in the dense thickets, Goa leads the file on to the sand bar, passing in front of the elephant, and a bit close. When the bull begins to flap its ears and paw the ground, Brian and Goa, exchanging guns, also exchange an old-days grin of recognition; they have met this situation many times before. Brian says later that they must have killed three hundred elephants together (of the estimated thirteen hundred he has executed in his lifetime), almost all of them on control work in Tanzania; it has been many years since he took one on license. Between one and three thousand elephants were destroyed annually in control operations throughout his twenty-three years in the Game Department, and this was only a fraction of the total number taken here in the southeast.

The bull does not yet have our scent, nor does he see what makes him nervous, although Brian claims that elephants' sight is better than people imagine; perhaps he has heard us, or perhaps what has come to him is some subtle shift in atmosphere that affects his sense of elephant well-being. Testing the air with lifted trunk, he steps down on to the river bed, then swings around to dig a hole under the bank. When the water wells up, he picks up a trunkful and hurls it overhead, so that it falls with a fine *splat* upon his back; he sprays himself behind his ears and under his belly. In the process he develops an erection, but the lumbering penis is gingerly, first lowering a little like a boom, then bobbing upward in alarm as its owner moves out over hot sand. Nearing the river at a point downwind of us, the elephant stops short, then turns toward us, trunk held high; the gland takes cover as he wheels away and hurries into the bush.

A big pan not far away up the east bank of the river has always been a famous place for elephant and game; in other days, this pan held a large pool that was permanent home to crocodiles and hippos, but the hippos eventually wore their runway to the river so deep that at last the whole marsh emptied out, obliging these two species to gain a living elsewhere. It was here at Likale, Brian says – and he points to an open wooded hill on the south side – that he saw the greatest and most splendid kudu of his life. On the north side, sixteen elephants feed on the green carpet of an *mbuga*, and a number of others are in sight, including two that cross the river to join those on the western bank. In the dry pan, one hundred buffalo stand in a compact herd, with nearly that number of impala, thirty wart hog, a band of kongoni with a new calf, a zebra herd at the far wood edge where had been seen the great kudu of yore, and a quorum of the elongated birds that stand about at the water's edge the world over.

Brian looks about him, saying, "I feel as if I was just here the other day. Nothing seems to have changed very much." I couldn't make out whether he was glad or sorry.

Having always been wary of returning to any wild place that has meant a lot to me in case it might have changed, and not for the better, I ask him why he has been so anxious to return to the Selous. At this, he glares at me, defensive. "Who said *I* was anxious to return here? Tom Arnold tell you that?" When I say nothing, but just meet his gaze, he comes off it with a sheepish smile. "No," he says quietly. "I get a lot of pleasure out of being here. See how the game is doing. Visit the old places. See my old staff." He glances at Goa, then blurts out, "The Selous is home to me, you see; it's the only place on earth where I feel I belong." Asked if he wanted to be buried here, as Ionides had been, he answers stiffly, "Don't give it much thought. Doesn't pay to be morbid. Don't expect I care very much *what* they do with me after I'm dead."

We wade the river and make camp just opposite the pan. To the east, the land rises to steep red escarpments; these broad valleys with steep cliffs may well be indications of ancient fault lines of the Great Rift that runs south from the Red Sea to the blue Zambezi. On all sides is an airy view of the Old Africa, and I am delighted that Brian wished to camp here and sorry to see the look of discouragement upon his face. "I'd been looking forward to this place," he says, "and it's the best we've seen, yet it just doesn't compare with the way it was. That buffalo herd used to be three or four times that size! And we have yet to see a single elephant with decent ivory!" It was just about this time last year that he and Arnold had flown up the Mbarangandu on a reconnaissance for this safari, "and there was a lot of game here, a *lot* of game, and all the way along: I don't believe we were ever out of sight of elephant." I protest that even a poaching epidemic, of which we have found no sign at all, could not have wiped out so many elephants in a single year, and at this he brightens; we should not forget, he reminds me, that the rains were very late this year, while last year they were normal. "We're well into September, and the Mbarangandu still looks as it normally does by late June or early July! The dry season is two months behind schedule, and the animals are still scattered. There's no need for them to concentrate along the rivers. There's water everywhere out in the bush, you've seen it for yourself!" For a moment Brian's voice is elated again, as if this familiar litany has cleared his doubts, yet, as I am beginning to perceive, a part of him has no wish to be consoled, for a moment later he is glum again. "I suppose it's a mistake to revisit a place you loved, to make this sort of sentimental journey. I haven't made a foot safari since 1963 – everything was Land Rovers after that – and a place can change a lot in sixteen years."

I repeat my arguments, to reassure myself as well as him: the scarcity of big bulls with large tusks is bothersome, of course, but it is simply not possible that the thousands of elephants observed from the air in 1976 and 1978 could have been killed off by poachers or even by plague, since on this trek we have found not one dead elephant. Also, the fact that the groups were so small was strong evidence that they were not being harassed. And wasn't there also a certain negative reassurance in the case of the rhino? Regularly along our way we have come upon fresh rhino scrapes, and since the rhinoceros regularly returns to the same place to defecate, almost all of these scrapes must represent different animals: yet the pair this morning were the first ones we have actually seen on our safari.

In the 1960s a number of rhino were killed because of a notion among Orientals that the compacted erect hair of the rhino "horn" was a cure for impotence and certain fevers: the accelerated rhino slaughter of today has come about because rich Arabs of the Middle East have made a fashion of daggers with rhino horn handles, for which they are willing to pay over six thousand dollars apiece. The fad or fetish for these phallic daggers has jumped the already very high price of horn up to five thousand dollars per kilo in Hong Kong – the worthless stuff commands more than pure gold – and unless drastic measures are enforced, and soon, an ancient species may vanish from the earth millions of years before its time because of sexual insecurity in *Homo sapiens.*

Even ten years ago one could take for granted encounters with a few rhinoceros; these days it is a stroke of luck to see one. So far as it is known, black rhinos have been all but eliminated from Uganda. Kenya's recent population of fifteen to twenty thousand rhino has been reduced to between twelve and fifteen hundred. Since the early 1970s, the rhino in the Tsavo parks have declined from seven or eight thousand to one hundred and eighty; in the small Amboseli park, the decline is from fifty to ten. Rhino poaching has crossed the border from the Mara into the Serengeti, and the other important Tanzania parks – Ngorongoro, Manyara, Tarangire, Ruaha – have already lost at least three-quarters of their populations. In the Selous the most recent estimate of rhino numbers, made during the air survey of 1976, arrived at the figure of four to five thousand, which must be the last large healthy population of this species in the world.

In the late afternoon, I wade the river to observe the large impala herd, which is mostly engaged in group activity that seems to anticipate the rut; while the does amble back and forth, uninterested, the bucks all run about, tail flags held high and barking like sick baboons, pausing here and there for a quick skirmish, running on. A few elephant and buffalo still linger at the edges of the plain, and the thirty wart hogs are still present, avoiding one another's company, moving about on their front knees and snuffling into the earth.

Dark clouds and wind. At dusk, under the eastern bluffs where an elephant is throwing trunkfuls of fine dust into the air, a ghostly puff of light explodes, another, then another. I cannot see the elephant, only the

(Right) River confluence.

(Over) Water-buck.

dust that rises out of the shadow into the sunlight withdrawing up the hill. At dark a hyena whoops and another answers, for the clan is gathering, but their ululations are soon lost in a vast staccato racket, an unearthly din that sweeps in rhythmic waves up and down the river bars, rising and falling like the breath of earth – then silence, a shocked ringing silence, as if the night hunters have all turned to hear this noise. Somewhere out there on the strand, I think, a frog has been taken by a heron; my mind's eye sees the long bill glint in the dim starlight, the pallor of the sticky kicking legs, the gulp and shudder of the feathered throat. The frog's squeak pierces the racket of its neighbors, which go mute. But soon an unwary one, perhaps newly emerged from its niche under the bank, tries out its overwhelming need to sing out in ratcheting chirp; another answers, then millions hurl their voices at the stars. The world resounds until the frogs' own ears are ringing, until all identity is lost in a bug-eyed cosmic ecstasy of frog song. In an hour or two, as the night deepens, the singing impulse dies, leaving the singers limp, perhaps dimly bewildered; remembering danger, they push slowly at the earth with long damp toes and fingers, edging backward into their clefts and crannies, pale chins pulsing.

Toward midnight I am awakened by a bellow, a single long agonized groan; a buffalo has cried out, then fallen silent. Perhaps something is killing it, perhaps a lion's jaws have closed over its muzzle, but I hear no lion, now or later. At daybreak a bird call strange to me rings out three times and then is gone, a bird I shall never identify, not on this safari or in this life. As a tropic sun rolls up on to the red cliffs across the river, setting fire to a high, solitary tree, the moon still shines through the winged piliostigma leaves behind the tent.

Kibaoni, or "Signboard", is a location on the Mbarangandu River where Ionides had caused a sign to be put up on the footpath between Liwale and Mahenge, reminding the Ngindo that they were in a game reserve and that it was forbidden to stray off the trail. A few miles short of Kibaoni we head west from the river, climbing gradually toward the red cliffs of a plateau. Though our destination is right there before us, Goa seems oddly disoriented and indecisive, wandering back and forth and tending into the wrong paths, until finally Brian stops him and explains the route. To me he says, "He knew where he was going, but never having been there before he had no sense of it. That's why he was wandering like that."

On a grassy hillside of small trees, a burst of terminalia saplings has sprouted out of a rhino scatter on the path, and there is lion spoor. Then Goa's hand is up: he points. A large dark animal stands in the high grass below the ridge line. With binoculars I pick out four more sable, lying in the copper-colored grass of the ridge summit; they do not look at us but face eastward, over a broad sweep of the river, two miles away. Then

(Left) Violet-tipped courser. (Preceding page) Monitor lizard.

another big male, long horns taut, is standing up and staring at us, his shiny black hide set off by the white belly, chestnut brow abristle with morning light above the twin blazes of his face. Apparently he gives the alarm, for now another bull jumps up, and then another. The bulls regard us for a little while before leading the herd away along the ridge line and down on the far side. A dozen animals cross the sky, including a very young calf, and probably as many more never emerged from the high grass but simply withdrew down the north side of the ridge. The two lead bulls watch the others go before moving up through a small glade of silver trees at a slow canter, turning against the sky to stare again, then vanishing from view.

"They're incredibly tame up here," Brian remarks. "Don't know why this herd pushed off so soon. I've walked up to within twenty yards of sable down in this part of the country."

At the foot of the plateau, heart-shaped prints of sable ascend a slide where elephants have broken down the small steep cliff to make a pathway up and down the escarpment; in precipitous places, an elephant may sit back on its haunches and brace all four feet out in front to slow down its descent, but sometimes it is killed or injured anyway, and I wonder if the first one to try it in this place might not be buried deep beneath all this red rubble.

On the flat tableland above the cliff is a stand of closed *miombo* woodland, and we have not penetrated it very far before we pass twin gouges of a rhino. Before long, light appears beyond the brachystegia, and the graceful dark trees open out upon an elliptical pan nearly a mile in length, ringed all the way round by the closed woods. "Goa has never been here," Brian murmurs, "and the Ngindo in this country don't seem to know about it, either. Even my esteemed predecessor never knew about this place. So far as I know, you're the only white man besides myself ever to see it." But as he speaks, he is scanning the lost pan: despite all the sign that we have seen, the well-worn paths that have beaten flat the woodland edges, the good clear water, there are no rhino, elephant, or sable, not a single animal of any kind. In a way, the emptiness makes the great pan more impressive – the stillness of the glittering water, the yellow water lilies and the tawny marsh grass, the circle of still trees that hide this lovely place from the outside world, the resounding silence and expectancy, as if the creatures of the earth's first morning might come two by two between the trees at any moment. A pair of jacanas stand in wait where the east wind stirs the floating vegetation, and a few swallows loop and flicker in eternal arabesques.

"*Pumzika,*" Brian says, and the porters, sullen, dump their loads; they seem to wonder why we have come all the way up here. Even Kazungu, who was mildly rebuked early this morning for letting the porters eat up all the food, has curdled a little in his attitude for the first time on our safari. Not wishing to talk, Brian goes off a little way and sits down with his back against a tree. Soon the wind dies. A scaly-throated honeyguide comes to the gray limbs overhead, calling out to us to follow; otherwise the woodland is dead still. Kazungu, who knows perfectly well that tea and porridge are expected after the first trek of the morning, is sitting himself down, doing nothing about a fire, until finally Brian must call out to him, "Is our tea ready?" To cheer him up, Brian adds, *"Tu na tupa macho yetu kambini,"* "We are throwing our eyes toward our camp" – in other words, From here we are starting home. Kazungu grins, and the porters look pleased, too; they have enjoyed themselves despite their labor, but now they are ready to return. Only Goa shows no elation; he is content with whatever Bwana Niki decides.

Before leaving Mkangira the Africans were given enough biltong to last eight days, or even ten if they took a little care; after five days the biltong is all gone. To carry loads, the porters say, they need something stronger than rice and porridge, and as the beans are already in short supply, another animal will have to be shot. Brian dislikes the idea very much. "For one thing," he says, indicating the waiting trees, "I would hate to break all this silence with a bloody great noise." We can take guinea fowl, wart hog, buffalo, or impala, and the first choice is impala: most of the Ngindo are Muslims, which eliminates wart hog, one or two guinea fowl would not suffice, and to kill buffalo would be very wasteful. "Anyway," Brian says, "I don't want to shoot at a dangerous animal with that Game Department .458 that I've never fired; I doubt very much that it has ever been zeroed in. If somebody got gored out here by a wounded animal, we'd be in serious trouble. The nearest settlement where we might send for help is at Liwale, at least four days away across rough country. Probably send us back a couple of aspirin."

From this nameless pan on this nameless plateau we shall head back north to the sand river where two days ago we surprised the resting lions, then follow it north and east to the Mbarangandu, which we shall descend to Mkangira. We realize now that this pan has been our outward destination, that today we are turning back to the New Africa, to an orderly life, to "civilization". At this prospect, Brian's expression, so clear and youthful in recent days, visibly sours. He wonders aloud if our companions have been fighting. "Always women who do the fighting on safari, usually against one another. Probably get back and find some of them camped on the far side of the river!" He is joking, of course, and starts to smile as a sly, bad expression comes across his face. "Who knows *what* they've been up to back there!" he exclaims, warming to his fantasy. "Buggery and lesbianism, probably! I tell you, the human animal when he gets despondent is bloody bad news!" And this bad old bush rat, Mister Meat, with his unshaven jaw and bad smile and loose teeth, removes his cigarette holder and laughs himself red in the face, and I laugh with him, not so much at what he has said but because of the infectious gusto of his cynicism.

Brian sighs, gazing about him. "In Africa, out in the bush, man is still a part of nature, and what he does is mostly for the better. It is only where the bloody Western civilization has come in that everything is spoiled." Brian has a poor opinion of *Homo sapiens* and his ambitions, but in the main he is amused by human folly, not made gloomy. Now he grins. "I told Melva once that human beings were the dirtiest and most destructive animals on the face of the earth, and she took it personally. Women aren't really very logical in these matters. But now she understands my point of view." He heaves to his feet, and sets off into the woods without a backward glance at the hidden pan.

On an elephant slide on the north end of the plateau lies an old cracked tusk that has been there for many years — mute evidence that no man comes this way. Descending, we head north through rough and unburned country. Goa is setting his fires again; columns of thick smoke rise up behind us as we head north to Mto Bila Jina, the River Without Name.

Coming down off the Luwegu side of the high plateau, this big stream that becomes a sand river in the dry season tends to the northward, bending close to the Luwegu. Eventually it curves away toward the northeast as a tributary to the Mbarangandu, an avenue of sand perhaps forty feet across, under steep banks lined on both sides by big trees; even now, deep in the dry season, there is clear water not two feet beneath its hot white surface. Because of the river's serpentine course, we do not walk along its bed but cut cross-country between bends, holding to our northerly direction, and in the early afternoon we break the journey at a rank meadow spring on an open hillside. When the day cools a little, we continue onward, skirting an occasional elephant, hearing the bark of an occasional Sykes monkey or baboon, crossing and recrossing the River Without Name. All the while, Brian is looking for impala, but there is no sign of impala, or of wart hog, or of guineas, only the baked savannas and dry hillsides and open woodlands of high grass.

Toward sunset, where the river rounds a bend, three buffalo bulls stand together at the end of a long stretch of clean white sand. Brian is footsore and discouraged and irritable, and he knows that for the first time on our safari, the morale of the Africans is precarious due to the real or imagined need of meat. Earlier he had said that in shooting a buffalo too much of the meat would be wasted, even if two porters gave their loads to the others in order to lug all the meat that they could carry, but now he decides without further ado to execute one of the bulls. Leaving the rest of us behind, he stalks with Goa to a point on the stream bank not twenty yards from the three buffalo below, a point-blank range from which he is sure to drop the animal with the first bullet. At the crack of the rifle, the buffalo sags down upon the sand with the windy groan of death, and in the echo of the shot, the Africans laugh and clap their hands together. No second shot is needed, and when we come up, I congratulate Nicholson on killing the buffalo with such dispatch; he refuses the small comfort I have offered. "I hate doing that, it really depresses me," he says. "But we were getting into a serious food situation with these porters. Karen and Rick have never been on a real foot safari, where beans are the staple; they gave us twelve kilos when what was required was a whole bloody sack." He swears to himself, restless, unable to make his peace with it; I had not suspected that he would be so upset. After all, this man has killed thousands of animals, and no doubt hundreds of buffalo among them; the buffalo is not an endangered species, and these three bulls may have passed the reproductive age since they had wandered off from the large herds. Even the "waste" will not be seen as such by the carnivores and vultures that will reduce this buffalo to bones in a few days. It is not the waste that bothers him, but the intrusion by man into the "heart of the Selous" which was symbolized by that isolated shot.

Already Goa is preparing for the butchering, cutting dense branches of the dark green adina trees to lay as a meat rack on the white sand beside the carcass. Despite the hard day they have had in the dry hills, the Africans are inspired with new resolve by the dead buffalo. "We'll have to camp here through tomorrow so they can smoke that meat; they'll start tonight. And they need the rest. That's one reason I decided to shoot that buffalo and have done with the food problem; there's good water right here, we can camp next to the animal, they don't have to lug all that wet meat out of the woods."

The inert dark mass lies sprawled on the white sand, tongue lolling, as ticks and flies crawl over its thin belly hair and testes. In sunset light, Abdallah cuts its throat, and the thick blood pours away into the sand, and still Brian Nicholson does not stop talking. "These buggers can gorge themselves on meat tonight, and with my blessing, and all day tomorrow, too. And the next day, one man, maybe two, is going to carry nothing but meat; the rest will have to manage the extra loads. Because I'm not shooting anything again." I have the feeling he may still be talking to himself as he goes off down the stream bed for a wash.

On the damp sand just beneath the adina and tamarind grove where we will camp, Kazungu has paused a moment in his digging, as if hypnotized by the upwelling of clear water. Beside him are fresh tracks of both leopard and lion, and as a yellow moon rises in the east to shine through the tamarind's feather branches, a leopard makes its coughing grunt not far downriver. Soon lions are roaring, no more than a mile away. Brian says that because of their poor sense of smell, lions usually depend on vultures to locate carrion for them, and that he'd seen hyena using vultures the same way, going along for a few hundred yards before cocking their heads to locate the spiral of dark birds, then trotting on again. Though he doubts that lion would find the carcass, hyena or leopard might come in this evening, under the big moon.

That evening I ask Brian if he would ever consider returning to the Selous were he given a free hand to reconstitute it. "You never know," he says, after a long moment. "Don't want to burn all my bridges behind me. But I worked for next to nothing all those years; they can't expect me to do that again. Don't want to wind up on the dole in some little charity hole in the U.K." – and here he looks up at me, genuinely horrified. "Oh God, how I'd hate *that!*" he says, and I believe him. There is nothing inauthentic about Brian Nicholson's self-sufficiency and independence, evolved out of hard circumstance very early in his life and reinforced at the age of nineteen when he banished himself to the wilds of Tanganyika. Unlike Rick Bonham (and unlike Philip Nicholson, the only son of a legendary warden of the great Selous) Brian has no romantic heritage in East Africa, or even a strong family to fall back on; neither did he ever have the celebrity enjoyed by the wardens of the great tourist parks, such as Bill Woodley and David Sheldrick and Myles Turner, the ones taken up by the shiny people who made East Africa so fashionable throughout the sixties. Not that he has complained of this, or even mentioned it; he has no self-pity, although here and there one comes upon a hair of bitterness. One day when he lost a filling, I told him he'd get no sympathy from me, not after his hard-hearted response to my shattered tooth back there on the Luwegu. This teasing was meant as the only sort of concern he could permit, but Brian failed to smile, saying coolly instead, "Unlike you, I *expect* no sympathy." There was a certain truth in this, but perhaps the remark revealed more truth than he had intended.

"The Selous ought to be set up under its own authority," Brian is saying, "financing itself and administering itself, not vulnerable to people who aren't really interested. That was the trouble down here when I left – lack of real interest. Now Costy Mlay, he was sent down here from Mweka, and he quickly understood the problems and saw the potential. Costy's the exception; he *was* interested, and he's *still* interested, even though he is no longer with the Game Department. Costy's very bright, and he's not a politician.

"To lose the Selous now would be such a dreadful waste, and especially when you realize that everything is present that is needed to administer it efficiently, all of the groundwork has already been done! For example, all the road alignment – when I was here, we were operating over three thousand miles of dry season tracks. And the placement and grading of the airstrips – that's done, too. They just have to be cleaned up again. There's even the nucleus of a good staff – the old game scouts who still know the country, who could train up a new corps of men, set up patrol posts. That's the first thing I would do if I ever came back here, get my old staff together to train up good new people, like some of these young porters we have here now. I'd try to persuade Damien Madogo to come back, and I'd get hold of Alan Rees. Rees was my right-hand man all the way through, he was principal game warden for the Western

Selous – held the same official rank as I did. No story of the Selous would be complete without mentioning Alan; he was a fine hunter and a fine warden, conscientious and patient and very good with his staff, and in addition, he's a first-class naturalist; he just loves being out in the bush, rains and all, and his wife is the same way.

"The senior warden up there in Dar, Fred Lwezaula – he's all right, too. He's doing the best he can, considering the fact that nobody up there in Dar appreciates what they've got down here. There aren't many people in the government who even know where the place is! I don't think it's ever been brought home to them that the Selous is unique. It is not just a big empty part of the ordinary monotonous *miombo* country that takes up most of southern Tanzania; it's well watered, it's vast, it's almost entirely surrounded by sparsely populated country, it's an ecological unit – or *several* ecological units, as Alan Rodgers says – and there's nothing like it in East Africa. But the Selous *has* to be self-supporting if it is going to withstand all the demands for land and timber and the like that are bound to come; we proved it could support itself very easily, and build up the country's foreign reserves as well, and *still* remain the greatest game area in the world." The Warden paused for breath, then concluded quietly, "If only for economic reasons, they owe it to the future of their country to see to it that this place doesn't disappear, because it's very precious, and it is *unique* – I can't say it too often! There's nothing like it in East Africa!"

Both lion and leopard cough and roar intermittently throughout the night, but at daybreak there are no fresh tracks around the puddle of congealed blood, the pile of half-digested grass stripped from the gut, the sprawl of entrails, the mat-haired head with the thick twisted white tongue. Overhead, the moon is still high in the west, and shimmering green parrots, sweeping like blowing leaves through the river trees, chatter and squeal in the strange moonset of the African sunrise.

(Top left) Abdallah. (Top right) Goa.
(Bottom left) Davvid Endo Nitu. (Bottom right) Mata.

XI

By the time the buffalo was butchered it was late last night, and the Africans were exhausted; the meat was heaped in a big pile by the kitchen fire to discourage theft by passing carnivores, and this morning, under Goa's direction, Mata and Abdallah are cutting heavy Y-shaped posts and setting them into the ground. The posts will support a rack of strong green saplings, and under the rack a slow fire will be tended that will keep the meat enveloped in thick smoke. "This is a day they will always remember," Brian comments. "Down here in southern Tanzania, they have no livestock at all except a few goats and chickens; the poorer ones get hardly any meat. So this is a unique experience for most of them, perhaps the one time they will ever have it – the day on which they were actually paid to sit around and eat all the meat they could hold. Of course that used to happen with the elephants I had to shoot, but those people weren't paid for it, as these are." He describes how in the old days, traveling light on elephant control, he would camp next to the killed animal in this same way, living exclusively on elephant kidney and sweet tea and rice cooked in advance and packed tight into a sock – a clean sock, mind, he adds, with the trace of a smile.

I take advantage of a day in camp to go off by myself and look for birds, walking alone up the sand river. By the dead buffalo, a vague cold smell of turning meat is mixed in a repellent way with fleeting sweet whiffs of bush orange, but soon there is only the faint mildew smell of the haze of algaes on the damp sand, which everywhere is cut by the tracks of animals; crisscrossing the marks of eland, kongoni, bushbuck, hippo, buffalo, and elephant are myriad patterns of unknown small creatures, and also the round pugs of lion and leopard. Since I am barefoot, it would be difficult to circle through the bush if something came down into the river bed and cut me off from camp: I listen for the crack of limbs, the buffalo puff, the rhino *chuff*. Brian had recommended the services of Goa and his .458, but Brian himself has a poor opinion of Goa's marksmanship, and to observe birds with an armed escort —! Anyway, I wanted to get off by myself.

Walking alone is not the same as trudging behind guns; one stays alert. And although this is not the first time on this foot safari that I hear the wind thrash in the borassus palms, the moaning of wild bees, it is perhaps the first time that I listen. Walking upstream, I am shadowed for a while by a violet-crested turaco which moves with red flares of big silent wings from tree to tree, then hurries squirrel-like along the limbs, the better to peer out at me, all the while imagining itself unseen; occasionally it utters a loud hollow laugh that trails off finally into gloomy silence, as if to say, "Man, if you only knew . . ." A large patch of blue acanth flowers on the bank is shared by the variable sunbird and the little bee-eater, and when I pause to watch the sunbird, a tropical boubou climbs out of a nearby bush and utters its startling bell notes at close range as its mate duets it from a nearby tree, then unravels the beauty it has just created with a whole run of froggish croaks that cause an ecstatic pumping of its black-and-cream-colored boubou being. Brown-headed parrots and a beautiful green pigeon climb about in a kigelia, which has also attracted a scarlet-chested sunbird. Cinnamon-breasted and golden buntings flit in separate small flocks across the river, and a pair of golden orioles skulks in a bush; ordinarily these shy birds frequent the tree tops. At a rock pool, perhaps a mile upstream, I watch striped kingfishers and a white-breasted cuckoo-shrike, and listen to a bird high in the canopy that I have never heard before and cannot see; its single note is a loud and clear sad *paow!* Circling it, waiting, listening, I am rewarded at last with the sight of a lifetime species, the pied barbet.

At camp, toward noon, the vultures are already gathering: fifteen griffons have sighted the buffalo carcass and are circling high overhead. But soon they have dispersed again, after swooping in low for a hard look; though nothing threatens vultures, they are wary birds, and too many humans for their liking come and go around our camp less than fifty yards away. There is a lion kill not far downriver, to judge from the resounding noise heard in the night; perhaps the vultures have gone there to clean things up.

In early afternoon, over the river trees, heavy rain clouds loom on the east wind. Then a light rain falls, and the returning griffons, accumulating in dead silence, fill bare limbs back in the forest with dark bird-like growths; at some silent signal, half a hundred come in boldly on long glides, feet extended; they strip the buffalo dry and clean within an hour.

By late afternoon the smoked meat is shrunk down to black curled twists of leather. The Africans take turns tending it, and the others sit upright in a circle under the tamarind: the small-faced Mata, and tall Amede, and the small man with the child's wide-eyed face who is called Shamu, and the squealing Abdallah with the squint, and Saidi Kalambo in his big hair and huge blue boxer shorts, and the heavy boy in the red shirt who is called "Davvid" although he is a Mohammedan – "*Davvid* Endo Nitu," he insists firmly. And Kazungu says, "I also have a foreign name of 'Stephen', but now I am proud of using my African name."

Only Goa does not join in as the others laugh and tell one another stories. Mzee Goa, as the Ngindo call him, Old Man Goa, lies flat out on a piece of canvas, making the most of his day of rest, staring up through the dark green leaves of adina and tamarind at the blue sky. On the tamarind bark over his head, a large agama lizard is pressing up and down in agitation, and not far away, on a broken elbow of a piliostigma, a tree squirrel sits calmly, observing the human camp. Across the sandy soil near my own feet, where I sit on my campaign cot before our tent, a blue-black hornet, hard tail flickering, is dragging a dying spider twice its size toward some dank hole where it can be sealed in with the hornet's eggs.

At dark, the Africans move up close to the kitchen fire, which dances and flickers on the naked skin of dark chests and arms; behind them rises a stand of tall pale grass, higher than their heads, and beside them, on a bed of adina fronds, lies the big pile of dried meat. In more places than one, I think, along the northern borders of the Game Reserve, groups of young Africans such as these must be smoking their poached meat around bush fires very much like this one.

Since his Swahili is excellent and since he enjoys the jokes as much as they do, Brian teases the Africans a lot, and the younger ones joke back at him, as much for their own amusement as for his. Like Kazungu, they are good-natured, never impertinent, there is no air of aggression in their merriment. Yet I notice that Goa, though he laughs sometimes at Bwana Niki's jokes, never volunteers a sally of his own; his is the last of the colonial generation. And the Warden retains the colonial manner with the Africans, giving orders in short and peremptory tones. Except in calling out a name – Kazungu! Goa! (he doesn't know the porters' names – they are all "Bwana") – he never raises his voice; intentionally or otherwise, he usually speaks in such low tones that Kazungu or Goa must come up very close to hear him, seating themselves on the ground before his feet. On the other hand – unlike Ionides – he is never abusive or sarcastic, never shows anger even when exasperated, and takes the time to chat with them and make them laugh, though he is embarrassed when I ask him to translate one of the jokes. "Got to keep them simple, you know," he says, reverting to his old White Highlands manner. This time he has told the porters that if they keep on eating up the white man's food, they will grow very pale and their hair will straighten, and though a Mombasa Kenyan like Kazungu is too sophisticated for this jest, the Tanzanian country boys enjoy it.

I suspect Brian Nicholson of liking Africans, despite all the conventional prejudices he displays. If I had any doubts about this, most of them would be resolved by the evidence of my own eyes and ears that Africans, from the simplest of these porters to an urbane, well-educated man such as Costy Mlay, seem fond of Brian, whose fluent Swahili must convey subtleties of understanding and even concern that are absent in most whites. Unlike Ionides, Brian was born in East Africa and has known and worked with Africans since he was a child. It is true that he has usually been in a superior's position with these people, which goes a long way to explain his preference for country Africans over those in the city, and no doubt he would agree with Ionides that Western civilization has reduced many first-rate Africans to third-rate Europeans; I no longer bother to point out that, forced to adapt suddenly to an African culture, a first-rate European would certainly be thought of as a third-rate African, at least for the first few hundred years.

Kazungu, who can speak some English, is teaching me a little Swahili, and Brian teases him, saying that Kenyans don't really know how to speak it. "The Brits have really buggered up the Swahili language. Say *Jambo* instead of *Hujambu* just for a start. And then they *answer* '*Jambo*', which is all wrong, too. Once asked an African up there, '*Ume elewa?*' which means, Do you understand? and he became frightfully offended – thought I was accusing him of being drunk. To be drunk is *kulewa* – not the same sound at all." And Kazungu laughs, nodding his head; he freely admits that these young Tanzanians speak better Swahili than he does, since for them it is not a lingua franca but a local tongue.

As camp cook and a man with a bit of English, not to mention experience of such cities as Nairobi and Mombasa, Kazungu has prestige among the porters. But he never abuses his position, in part because he is outnumbered six to one. This evening, in the only unpleasant tone I have heard in the past fortnight, Davvid Endo Nitu is informing Stephen Kazungu Joma that if the arrogant Kenyans don't behave themselves, the Tanzanian Army will march through their country as it did through Uganda, to teach them a lesson once and for all.

Brian doesn't think much of my argument that the temporary anarchy in Uganda is probably worth it to get Idi Amin out of power, although I back up my position by relating a story told to me by Maria's brother, now a doctor in Australia, who had done his colonial service in Kampala. One of Peter Eckhart's teammates on the rugby team was a young African giant who was already heavyweight boxing champion of Uganda and a marvelous athlete, Peter said, immensely powerful and very fast for his great size; what struck Peter, who is not liberal in any way and doubtless shares the political outlook of Brian Nicholson, was this man's gentleness and consideration toward all the white players and absolute ferocity toward other blacks – "murderous", to use Dr. Eckhart's word, and a good word, too, in the light of later events. And then there was the warning that the British left the new leaders of Uganda when Independence came, concerning a young lieutenant named Idi Amin who was ruthlessly exterminating recalcitrant tribesmen up in Karamojong, how they were sure to have trouble with this fellow if they didn't bring him under control immediately . . .

And Brian shrugs, not because what I am saying is not true but because such excess is only to be expected on this continent: the bloody display of power is an African tradition. "Why was Amin any worse than that emperor up in Central African Republic, or the one in Guinea? Butchering people right and left, and nobody paid any attention." I cannot deny this. Who had concerned themselves about the deaths of the thousands massacred in Guinea by President Macias Nguema Biyogo? What could his Belgian friends reveal about political oppression under Colonel Joseph Mobutu, the Western puppet still catering to European interests in Zaire? And hadn't the French cynically supported the "Emperor" Jean-Bedel Bokassa of the Central African Republic despite the well-documented knowledge that he had personally taken part in the massacre of two hundred schoolchildren?

We are edgy on this subject, and I am considering how to change it when I notice the bottle of whiskey that sits like a reproach beside my cot. Since wine and beer were too bulky and heavy, this bottle has been brought along for me instead, but in the well-being and exhilaration of the foot safari, I have felt no need for it; on the other hand I am mildly embarrassed, since it has been an extra weight in someone's load. Thoughtlessly I suggest that it be given to the staff, to celebrate when we reach the Mbarangandu, and Brian snaps, "Absolutely not! You Americans ruined your own natives with alcohol; leave ours alone!" I know he has taken advantage of an opening, and I know why, but I manage to repress a sharp retort about the echo of colonialism in that "your" and "ours"; his choice of words does not change the fact that he is right.

Tonight we have buffalo kidney to go with our rice and beans, and while we eat Brian describes how he had once been hurt by a buffalo that had been wounded by a "big-game hunter" from Dayton, Ohio. "That was in the Maasai Mara, about 1949, when I was learning the hunting safari business as assistant to a chap named Geoff Lawrence-Brown. We had followed up this fellow's wounded buffalo and it jumped out of my side of the thicket, right on top of me. I got two rounds off into this dark blur, and it went down, but not before it caught me with the outside of its horn and sent me flying. I was pretty badly banged up."

I ask his opinion of these "big-game hunters", and the Warden grunts. "Don't know much about these people, really. Only stayed in that business long enough to qualify for my professional hunter's license. I always had my eye on the Game Department, you see. But I will say that most of the professional hunters I knew had contempt for a lot of their clients. Some tried to find nice things to say, and were loyal to their clients whether they liked them or not, but there aren't many professionals of that caliber any more. And a number of clients simply weren't physically fit enough to track their animals, even if they had nerve enough to leave their Land Rovers. Sometimes they shot from the windows, or rested their rifles on the bonnet of the car; many animals are unafraid of cars, so they had them at point-blank range. Sometimes the hunter and his client got out of the car on the far side from the animal, using the car as a hide; as soon as the car moved out of the way, they shot, and later it was claimed that they hunted the animal on foot. The client forgets what he wants to forget by the time he gets back home.

"A lot of these clients are first-class people, of course, but others are just big drinkers and big talkers, very childish men. What amazes me is how they worship their professional hunters, look at them like gods. I've seen grown men who seem to have good sense in every other way – *must* be good businessmen, at least, if they can afford that kind of safari – go all to pieces over one of these professionals who might just be a bloody idiot, and often is. Some of them are good men, good hunters and good conservationists, but too many are in it part-time, looking for quick money, or perhaps they're farmers who've shot a buffalo or two – short on experience and long on bullshit.

"One damned fella shot a rare cheetah down here, mistook it for a leopard – how could you mistake a cheetah for a leopard? The way they look, the way they move – why, they're not alike in any way! Another one told me he'd seen lesser kudu here, and dik-dik. Well, he hadn't.

"One hunter had trouble qualifying; there was no one who would vouch for him. He claimed to be a friend of mine, and used me as a reference when he applied for his license; probably thought they wouldn't check, since I was so far from Nairobi. I wrote back to say I'd never heard of him, which was the truth, but somehow he got his license anyway. And there was another one who claimed he knew me, too, one of these birds with a leopard-skin hat band and so many elephant-hair bracelets he could hardly lift his arms, you know the type. Told Billy Woodley he'd worked with me on elephant control down along the Ruvuma. Well, one day Billy and I were going through Namanga, and Billy introduced us, or rather, he said, 'I don't have to introduce you two, since you already know each other.' For some reason, the subject of our days together down on the Ruvuma never came up.

"Here in the Selous, we didn't care much how these people got their animals; if the professional did most of the killing for his client, as was sometimes the case – the client fires, and he never hears that second shot – it was more efficient and a lot more merciful. The point was that the Game Department needed the revenues, and every animal was paid for; a game scout went along with each safari to record any animals lost when wounded, since those counted on their limit, and to see that the limit was never exceeded. Each year we established a quota for each hunting block, according to what it could easily support, and we were very strict. I caught one German hunter who'd killed a rhino, although he had none on his quota, then tried to bribe one of my game scouts to keep him quiet; the scout came straight in and reported it, bringing the money. This German had also let an unqualified assistant take people out after buffalo, which was forbidden. I confiscated his license and told him he had forty-eight hours to get out of the Selous. The clients begged me, of course, but I just told them that they'd have to find themselves another hunter."

A cold clear morning. Well before daybreak, voices murmur and human figures move about, building up the fire to keep warm. A smell of carrion hangs heavy on the air, but the leopard, heard again last night, has not visited the buffalo, nor did the lions follow up our circling vultures. As for hyena, none have been heard since we left the Mbarangandu, nor are there hyena tracks in the sand river.

The sun-dried meat is packed into the loads; every man must help to carry it, since they mean to take it all. In the cold sunrise, the porters are quickly ready. As we depart, a stream of parrots in careening flight recaptures the sausage tree across the river; the fleeting human presence will lose significance with the last figure that passes out of sight in the dawn trees.

We are headed north again, into burned country, and the spurts of green grass in the black dust are sign that these fires preceded those we made on the way south; if this is the eastern edge of the large burn that we struck in the first days of our safari, then we are closer than we think to the Luwegu. In the bright grass the animals are everywhere, making outlandish sounds as we approach; the kongoni emit their nasal puffing snort, the zebra yap and whine like dogs, the impala make that peculiar sneezing bark. But the two buffalo that canter across our path are silent, the early red sun in the palm fronds glistening on their upraised nostrils, on the thick boss of the horns, the guard hairs down their spines, the flat bovine planes of their hind quarters.

At the edge of the plain, between thickets and karongas, Goa rounds a high bush and stops short; without turning around he hands the rifle back, as Brian and I stop short behind him.

In a growth of thin saplings, at extreme close quarters, stands a rhinoceros with a small calf at her side. The immense and ancient animal remains motionless and silent, even when the unwarned porters, coming up behind, gasp audibly and scatter backward to the nearest trees. Goa, Brian and I are also in retreat, backing off carefully and quietly, without quick motion: I am dead certain that the rhino is going to charge, it is only a matter of reaction time and selection of one dimly seen shadow, for we are much too deep into her space, too close to the small calf, to get away with it. But almost immediately a feeling comes, a knowing, rather, that the moment of danger, if it ever existed, is already past, and I stop where I am, in pure breathless awe of this protean life form, six hundred thousand centuries on earth.

In the morning sun, reflecting the soft light of shining leaves, this huge gray creature carved of stone is a thing magnificent, the ugliest and most beautiful life imaginable, and her sheep-sized calf, which stands backed up into her flank, staring with fierce intensity in the wrong direction, is of a truly marvelous young foolishness. Brian's voice comes softly, "Better back up, before she makes up her mind to rush at us," but I sense that he, too, knows that the danger has evaporated, and I linger a little longer where I am. There is no sound. Though her ears are high, the rhinoceros makes no move at all, there is no twitch of her loose hide, no swell or raising of the ribs, which are outlined in darker gray on the barrel flanks, as if holding her breath might render her invisible. The tiny eyes are hidden in the bags of skin, and though her head is high, extended toward us, the great hump of the shoulders rises higher still, higher even than the tips of those coarse dusty horns that are worth more than their weight in gold in the Levant. Just once, the big ears give a twitch; otherwise she remains motionless, as the two oxpeckers attending her squall uneasily, and a zebra yaps nervously back in the trees.

Then heavy blows of canvas wings dissolve the spell: an unseen griffon in the palm above flees the clacking fronds and, flying straight into the sun, goes up in fire. I rejoin the others. As we watch, the serene great beast settles backward inelegantly on her hind quarters, then lies down in the filtered shade to resume her rest, her young beside her.

We walk along a little way before I find my voice. "That was worth the whole safari," I say at last.

Brian nods. "Had to shoot one once that tossed a porter into a thorn bush and wouldn't give up, kept trying to get at him. But by and large, the rhino down in this part of the country have never given me much trouble." He turns his head and looks back at me over his shoulder. "Still, that's a lot closer than you want to get, especially with a gang of porters. If this was Tsavo —!" He rolls his eyes toward heaven. In an easing of the nerves, we burst out laughing, and the Africans, awe-struck until this moment, laugh as well: *kali* ("hot-tempered", dangerous) or not, a rhinoceros with new calf ten yards away was serious business!

Yet seeing the innocent beast lie down again, it was clear how simple it would be to shoot this near-blind creature that keeps so close to its home thickets, that has no enemies except this upright, evil-smelling

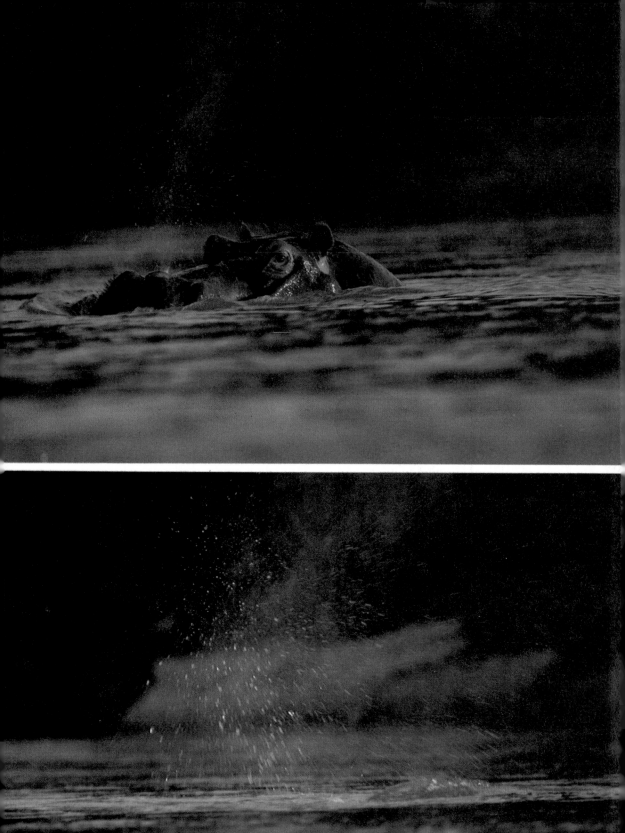

shadow, so recent in its ancient world, against which it has evolved no defense. Its rough prong of compacted hair would be hacked off with a panga and shoved into a gunny sack as the triumphant voice of man moved onward, leaving behind in the African silence the dead weight of the carcass, the end-product of millions of browsing, sun-filled mornings, as the dependent calf emerges from the thicket, and stands by dumbly to await the lion.

We head northeast into dry grassy hills. Big pink-lavender grass-hoppers rise and sail away on the hot wind, the burring of their flight as dry and scratchy as the long grass and the baked black rock, the hard red lateritic earth, the crust of Africa. To the west rise rough black escarpments, and beyond the escarpments an emptiness in the air, arising from the depression of the Luwegu's valley. Toward the southwest border of the Reserve shrouds of dull smoke ascend to the full fire clouds, all across the Mbarika Mountains.

The path descends into inland valleys of dry thorn scrub and long clinging strands of shrub combretum, then small sand rivers of sweet musky smells and cat-mint stink where crested guinea fowl, a shy forest species, run away cackling under the thickets. But there is no water, and because of their extra loads of meat and the strain caused by the encounter with the rhino, the porters are already tired before noon. To Brian's annoyance, they keep falling far behind. "It's very easy to get lost in bush like this," he mutters. "They must stay together."

Beneath borassus palms a small trickle of good water runs along the bed of a karonga, but under the red banks a juvenile elephant is swatting its legs with a fan of fresh-broken branches, and across the ditch at the wood's edge a cow moves back and forth in agitation; then a second cow with a smaller calf moves into view and disappears again. It is hard to tell how many elephants are here, or where they are, but the nearest among them are a lot too close. Uneasily, the porters set their loads down, all but the small wide-eyed Shamu, who stares astonished at the elephants like a little boy. Goa says sharply, "*Tua!*" (literally, "to land") – "Put your load down!" And Shamu does so as the second cow emerges part-way from the nearest thicket, ears flared out, and drives us back with a loud blare of warning. The confused calf in the karonga tries clumsily to climb the bank, and a third elephant, until now unseen – doubtless the mother of a calf we have not seen either – comes for us across the karonga. "*Kimbiya!*" Goa tells the porters. "Run!" The Ngindo scatter off into the bush, and the rest of us back up rapidly for the second time today. But the cow has stopped behind that wall of vines; we wait and listen while a boubou, startled out of its bush by the elephant, flies across the karonga and resumes its duet without a care.

The calf in the karonga is safely away, and the other calves are, too; we do not see the elephants again. Those in the bush right in front of us have simply vanished, so silently and so magically that even Brian can't

(Preceding page) Water-buck on the Mbarangandu delta.

[201]

quite believe it, and goes poking about in the bush in a gingerly way, just to make sure. Meanwhile, the porters have traveled so fast and so far that we have trouble reassembling them. "We lost track of ourselves," Mata admits, in the wonderful translation of Kazungu. We cross the karonga, drink warm water, and rest in the cool green shadow of an afzelia. Brian is pensive. "Have to be very careful with these animals, more careful than usual. I don't want to provoke a charge if I can help it. We're a long way from help, you know, if somebody gets hurt, and there's no track for bringing in a Land Rover. These emergencies can happen very fast, even when you see the animals and take pains with them, as we did here." He shakes his head. "What you don't want is to have one elephant cut off from the others. You try to make certain that they're all gone past before you push ahead, but sometimes in dense thicket like this, there's one that's slow or old that you don't see, and then there's trouble."

I am content that the Warden is being careful, and after all these encounters I have complete confidence in his nerve and expertise.

"One time, up on the Ruaha River – and by the way, we have more of the Ruaha right here in the Selous, about seventy miles of it, than they have in the Ruaha National Park – I had an elephant get in among my porters after I'd gone past. Chased one poor devil way off into the *bundu*, hot after him, you know, squealing like hell. Usually an elephant is quiet when it comes for you, at least until it's right on top of you; that's when it starts to yell. But this old cow yelled all the way, so we knew just where to follow. Kept coming on scraps of the porter's clothes – the turban he wrapped up on his head to cushion his load, then his shirt, his *shuka* – everything! He was running naked! But there was no sign of the body, so we kept on calling. After half an hour we reckoned that the poor chap must be a goner. We kept on calling, *Hoo!* – and suddenly, from a great distance, we heard, *Hoo!* The chap returned on his own two feet, all scratched and scared, but he was in one piece. Those clothes had saved him. Hadn't thought to shed them, they were just ripped off by the thorns, because with that elephant behind him he wasn't fussy about his route, he just went *moja kwa moja*, straight on through. But the clothes distracted that old cow just enough to slow her down. He never knew whether he'd outrun her or eluded her; he'd just kept running flat out, he said, until he realized that there was no more noise behind him."

We have hardly started out, toward three, when we run into more trouble. From a deep thicket comes a profound ominous mutter. Exasperated by the failure of the porters to keep up despite their hard experience this morning, the Warden begins a low muttering of his own. "What are these old cows so cross about?" he frets, cautioning the oncoming porters to be quiet. At that moment, the hidden elephant gives a loud and scary blare of warning close at hand – too close for the shot nerves of the Ngindo, who drop their loads and flee without further ado.

As Goa and Brian exchange guns, and Brian jams cartridges into the chamber, a Sykes monkey gives its loud *yowp* of alarm, and there comes a single sharp loud crack of breaking wood, then a dead silence. After a few minutes, when nothing happens, the porters are whistled for, and come in very quietly, one by one, grab up their loads, and flee again, loads to their chest. Not one was smiling. The porters are *choka sana*, very tired, from carrying the extra loads of meat and from the strain of all these encounters with big animals, and for the first time on the safari feel free to say so. "These elephants really made a pest of themselves today," Abdallah sighs as we pause to sip water from a ditch.

Our recent adventures with animals have broken down some of the formality between blacks and whites, the separation between their safari and ours. Kazungu tells me that he has never been on a foot safari before, nor have any of the young porters except Mata. Through the big voice of Kazungu I try to explain to the young Ngindo how precious this wild country is, where the water is clean and the game plentiful and even the dread rhinoceros is so peaceful. All the young porters nod fervently, saying "*Ndio, ndio, ndio,*" and Abdallah says, "It is good to see our country, and where the animals are staying," at which they all murmur "*Ndio!*" once again. It turns out that Abdallah is not an Ngindo but a Makonde, the coastal tribe now famous for wood sculpture. Although he seems no older than the rest, Abdallah is married and has one child, and Kazungu says that he himself has plans to marry, possibly next year. Kazungu has kept up his journal:

> About 9 a.m. we suddenly met a rhino. . . . To avoid being attacked by this rhino, we had to go slowly backward. I was very frightened. For a few minutes, nobody knew which way to go . . . This was the first time I saw a rhino face-to-face.
>
> After leaving the rhino we went right down to the river to make the porridge, then resumed our safari. The people were behaving very well, there was no trouble, they were more attentive now to what they were told, they were polite. . . . The safari was hard for people carrying loads on their heads and shoulders, but through good cooperation in a tough position, everybody was happy. They were tired, but still they were enjoying themselves. The only trouble they had was getting adjusted to hunger before the morning meal.
>
> As we were proceeding with our safari, we met some elephants resting. When they realized we were around they started screaming, and we were shocked, and ran away and climbed on the trees, leaving our belongings behind. I knew I would be all right because I stayed close to the person carrying the gun. Everybody ran away except Goa, Bwana Niki, Bwana Peter, and me.

Of the afternoon's encounter, Kazungu wrote:

> We could not see this elephant, could not see or hear anything with our ears and everyone was trembling. The elephant started screaming very highly, and this time the belongings were scattered around, and even I dropped my luggage and ran away. Later we started calling to see if we could find each other.
> After all this, we went down to the Mbarangandu.

Sand rivers, river thickets, green *mbuga* and dry black cotton marsh, grassy ridges and airy open woods, yellow and copper, red and bronze under blue sky – the *miombo* woods are bursting into multicolored leaf, well before the onset of the rains.

In mid-afternoon, along the edge of the wood, a female kudu steps out into bright sunlight. She is crossing one of the black granite platforms inset like monuments in the pale grass, so that even her delicate hooves are clearly visible. A second doe, already off the rock, awaits her. Then a magnificent bull kudu – all adult males of the greater kudu may be called "magnificent" – moves in the same slow dream-like step over the rock. The kudu are upwind and do not scent us, nor have they heard the sound of our approach. They fade into the woodland. But where the elephant path tends uphill toward the woods, the bull awaits us; this time he raises big pink ears as if to listen to the oriole in the canopy, then stops and turns and gazes at man for the length of a held breath, displaying all the white points of his face – the white muzzle and cheek stripe, the white chevron beneath the eyes, the ivory tips of the great lyrate horns. Then he wheels and canters up the ridge, disappearing in seconds, although a kongoni that pronks along behind him is visible for a long way into the woods.

It is late afternoon when white sand bars and the blue gleam of the Mbarangandu appear like a mirage in the parched valley, and near sunset when we reach the river plain. Near the river, Goa's sharp eye picks out an elephant skull deep in the thicket; he emerges carrying two tusks, which he will deliver to the Game Department. But the gunbearer cannot be expected to carry, and Abdallah and Mata take them away from Goa without grumbling and strap them on to their own loads, although each tusk weighs at least twenty-five pounds.

East of the river a dust devil arises, and one of the porters murmurs "*Moto*" – "Fire". Brian explains that the dust devil is actually a kind of thermal, upon which Goa, who almost never speaks except when spoken to, makes one of his rare utterances. As the younger Africans stare at the white men, wondering how we will take it, Mzee Goa says forcefully, "*No!*" Brian awaits him. "No!" Goa repeats in his deep voice. "That wind is made by a very big kind of snake. This snake lives in the Taita Hills, under the hill called Kasigan; this snake looks after the well-being of our hills. When it stirs, there are rock slides, and when it emerges, the spirits take it up into the sky in one of these spiral winds. The big snake is not visible, but if you listen carefully you can hear a kind of ringing."

Impala go bounding off in all directions, as if to spread the bad news of man's arrival, and in an harmonious broad bend of the river a group of kongoni on an evening outing walk sedately on the sand, escorting with ceremony a new and pretty calf, as if they could scarcely believe that it was theirs. Seeing the oncoming men they bounce away upriver, and because the calf still runs like an antelope, it keeps up with their odd progress, making long ground-gaining leaps like a small impala. But neither impala nor kongoni have run far before the flight impulse forsakes them and they turn their big eyes and big ears to take us in.

Beyond the kongoni five elephants, dusted the red color of the Tsavo elephants of Goa's youth, stand peacefully in the sun-bloodied water, and downriver a solitary bull feeds in a meadow just behind the bank. Like all the bulls that we have seen this is a young one, with small ivory, and reading my thoughts Brian says, "Where are the big bulls, Peter? I can't pretend to myself much longer that all those elephants seen from the air are back up in the thickets, or the big herds of buffalo, either; we've *been* back in the thickets, and they're just not there." I remind him that we had been lucky to see rhino, despite the abundant sign of rhino presence everywhere; it would have been simple to miss these enormous beasts entirely in a country as huge and wild as the Selous. Brian nods; he is not really upset. "Upriver from Kibaoni, now, where this river opens out in great flat plains, there's a lot of wild cane that elephants seek out in the dry season; perhaps *that's* where they all are. Perhaps they've all gone further south."

On the river sand, the footprints of ten men obscure the sinuous tail trace of a large crocodile, set off by hieroglyphs of odd long-toed feet. September sandpipers sweep up and down the bars, and a sand plover flutters forth in injury display from a nest depression somewhere near our path; the sandpipers are autumn migrants from the northern continents, but for the sand plover of austral Africa, it is now spring.

An open point set off by borassus palms, with a prospect of open plains and hills and broad bends of the river, and animals in sight in all directions – here was a camp site of the Old Africa that would have been chosen by those men of unknown color who left stone tools on the black granite, more than a thousand centuries ago. We bathe in the river, and afterward I fix myself a whiskey with fresh river water and sit propped up on my cot, on a low open bluff under borassus, gazing out across the sweep of sunset water to the green plain of Africa beyond; I feel tired, warm and easy, and awash with content. Kazungu brings good buffalo stew, and as the stars appear we listen to a leopard just over the river – big deep coughs, well-spaced and strangely violent in a way that the lion roar is not, followed by that rough cadence so like the sound of a ripsaw cutting wood. Perhaps the leopard is disturbed by the unfamiliar light of a fire across the river; perhaps, like that gentle rhino, like the tame antelopes of the southern Selous, it is only vaguely troubled by our intrusion, having had no experience of man.

"I reckon the leopard is about the wildest animal there is. He's keen in eyesight and hearing as well as sense of smell, and he can hide so well that you can't see him when he's right there next to you; that's why following up a wounded one is bloody dangerous. Back on the old farm in Subukia, when I was a kid, there were still a lot of leopard about, raiding the stock; I had a big skin on the wall of my room that scared me, I used to imagine awful encounters with leopards." In his disreputable green *shuka* and red sneakers Brian lies back on his cot, fresh-shaven, towel around his neck. "My stepfather was mauled by one. This leopard had been taking his pigs, and he got a shot at it and wounded it. Perhaps half the time a wounded elephant or buffalo, even a lion, will push off again when you follow it up, but not a leopard. A wounded leopard is going to come for you the very first time you bother him, every time; sometimes he doesn't run for the thicket at all but comes straight for you after the first shot. He's so bloody fast that you don't realize he's coming until he's already covered half the ground between, and he isn't a big target, either – if you manage to stop him before he's on you, you're bloody lucky. Anyway, my stepfather was a farmer, he hadn't much experience of the bush. When he went after that leopard he was torn apart, and later they had to amputate his leg. No antibiotics then, of course; he was lucky he survived the septicemia."

When I ask Brian if he liked his stepfather, he frowns. "My stepfather was a steady sort of chap, but I never knew him well enough to know how much I liked him," Brian says, slightly uncomfortable. "They send you away to school pretty early in East Africa; I was only about eight, and even when I managed to get home, he was always out on the farm all day." Brian raises his eyes and looks me over, as if considering whether to say what he says next. "Didn't know my real father at all. Had a farm at Eldoret where I was born, but lost it during the Depression. My parents divorced when I was three, and I was given my stepfather's name when my mother remarried and we moved down to Nakuru. My dad came down here to this country, it was Tanganyika then; his name was Dickinson. He died in South Africa in 1967, and the rest of them were scattered to South Africa and Australia and Rhodesia, all but my half-brother Mike, who's a first-class wildlife painter in Nairobi. Can't say I know Mike, really. My mother's in Johannesburg with my full brother; haven't seen her in the last twelve years. My brother's in business there. I can't say I know him, either; never did." After another pause, he shrugs. "Ionides was sort of a father, I suppose, though I was already nineteen when I first met him."

Considering his history, one can quite understand the susceptibility of this young recruit to the fierce Ionides, who was not only a legendary hunter but the despotic ruler of a vast, unknown domain. And the impression made was a profound one, to judge from apparent similarities in attitudes and statements, even style: a certain odd smile, a cocking of the head, an indifference to food, alcohol, and dress are all traits that were noticed in his mentor, and so is the defensive habit of trying to conceal enthusiasm or excitement in order to be prepared for disappointment. But Brian Nicholson, for all his self-imposed isolation, needs a woman and perhaps he needs children, and try as he may, he cannot achieve the coldness of Ionides or even his notorious "eccentricity", which by all accounts consisted mainly of cranky self-absorption and a rather childish defiance of conventional restraints such as might have been imposed at Rugby School. For example, Ionides felt constricted by socks, underwear, and neckties, and dispensed with these even on his rare visits to London.

"Iodine was thought 'eccentric' because he liked living alone off in the bush. He didn't need people. He had all those books he left to me, and he could quote from every one of them, and I reckon that was all the company he needed. They also decided he was 'eccentric' because he didn't bother much about his clothes, and because he liked snakes, and because he visited a witch-doctor and let himself get bit by a young cobra, to try to establish antibodies in case of snake bite. Not the least bit eccentric, really. I'd have been thought eccentric, too, if I hadn't been more conventional and got married.

"Iodine warned me about getting married, but I didn't listen; I had been out in the bush too long. Having small children, Melva rarely came out on safari in those early years; she scarcely set foot in the Reserve until after we moved up to Morogoro. She keeps the house beautifully, she's a lovely cook, she's good at gardens, and of course she saw to the three children. She helped me in many ways, of course, but she never became part of the work out here, she saw the Selous as a visitor might see it.

"You know, Peter, it's a funny thing," Brian continues, after a pause. "We saw so few people all those years, and now that we're back there in Nairobi, we don't see many people either; I don't even see very much of Billy Woodley. Mind you," he says, "I've been out of the Kenya wildlife scene since I came down here, and I've just been flying bloody airplanes since I went back." He grunts. "Tried to get Billy down here more than once. He always said the same damned thing that Myles Turner used to say: 'You've got a great bit of Africa down there, I'll have to come and have a look at it!' " The Warden shakes his head. "Neither of them ever came. It always amazes me how few people interested in wildlife have taken the trouble to come to the Selous."

Brian sighs. "It used to be that the day I left for overseas, I was already pining for Africa, not just the animals, you know, but the whole way of life out here. But since I've been up in Nairobi, I don't feel like

[207]

that any more; the way Nairobi is today, being there is like being overseas, and I don't mind leaving it at all."

Rick Bonham had once mentioned to me having seen some fine bronzes and old books of Africana that Ionides had given Nicholson; to Rick's surprise, Brian had said more than once that he wished to sell them and tonight, in fact, Brian asks me what I know about the prices of old books and bronzes, describing those he had been given by Ionides. He had long since got rid of a pair of vast tusks, 101 and 103 pounds, from a great trophy elephant that he had shot on license, long ago up on the Kilombero, and he had already sold off all but one of his fine guns, a Rigby .275 that he had kept aside for Philip. The guns had been sold quickly in the bitter year that he had quit the Game Department and returned to Kenya, and perhaps, I thought, he would sell Ionides's books in the same spirit; possibly this was a way to put behind him his bitterness about what was happening to the great kingdom of the elephants that he and Ionides had created.

Brian recalls that one of his old hunting rifles had gone to an assistant warden named Johnnie Hornstead, "one of the most generous fellas you would ever meet, and a fine mechanic, too. Old Johnnie was a hell of a mess – great big stout fellow with a big beard, used to go around barefoot all the time. One time there was someone out here, taping a television show, and this TV man said, 'Well, Mr. Hornstead, you must have a lot of time on your hands out here in the bush, may I ask what you do in your spare time?' 'Just mess around,' Johnnie informed him. 'Well, that's very interesting, Mr. Hornstead, but would you mind amplifying that a little? Can you tell us what it means to mess around?' 'Why, certainly,' says Johnnie, 'what I do, you see, what I do is, well, I just bugger about!' " Brian whooped with laughter at this memory of old Johnnie, shaking his head. "One day when Johnnie was drunk, he was just sitting there sprawled out, barefoot as usual, a lot of food dribbled down on to his beard, and his drink, too, and the hair on his head all standing up and filthy dirty, and all upset about something or other that I'd said to him on the subject of evolution. And finally he opens up one bleary eye, belching, you know, and he says, 'I don't care if you *are* my superior, I don't care if you give me the sack for it, I'm going to tell you something, Nicholson — you're socially unacceptable!' " And telling this story, Brian laughs so hard that tears fall from his eyes, and we set each other off, making such a noise that even the Africans stop talking, and in that instant I realize that for better or for worse this socially unacceptable man and I have become friends.

When we stop laughing, we are quiet for a while, listening to the passing of the river. Asked if he thinks our foot safari has been worth while, he nods, saying, "Frankly, Peter, I've enjoyed your company." I have certainly enjoyed his, and say so; I am pleased that both of us can acknowledge this without embarrassment, however anxious we might be

[208]

to change the subject as rapidly as possible. "I think this is the prettiest place we've made camp yet," Brian says finally, looking out over the gold-red of the river. "You know, I reckon I'm one of the very last people left who's done the real old African foot safari, staying out sometimes for months on end. Trekking with porters through remote, wild, wild bush like this, that hasn't changed a bit in three hundred years – that's not done in Africa any more. In Kenya, people just jump into their Land Rovers and minibuses and combis and away they go, but there really isn't any place left to go to. I saw the last of it up there twenty or thirty years ago, and the Selous is the last of it down here, make no mistake. That's why Rick Bonham is so excited. He's too young to have seen how Kenya was; this is the first time in his life that he's had a look at the real Africa. Wants to move right down here, set up a safari camp. Philip, too," he said. "When I was his age, I was already down here on elephant control, and of course he wants to do what I did, and he can't; for one thing, Philip, I keep telling him, you're the wrong color. And for a second thing, you can't be me." No more, I thought later, than Brian Nicholson could be C. J. P. Ionides, or than Ionides could be Frederick Courtenay Selous. Yet, different though they were, there was a certain continuity between these three unsocial animals, who had strayed out of the herd existence into a hunter's life that, as someone has written, was "lonely, poor, and great."

As he sits there barefoot in his *shuka*, freshly washed and his hair combed, and fit again after a fortnight in the bush, I have a glimpse of the young Bwana Kijana, come down to the Tanganyika Territories to join the Game Department on elephant control; in the strange half-light of sunset, the lines gone from his face and the gaze softened, he appears much as Ionides must have first seen him, as he must have appeared to the pretty Australian girl named Melva Peal, now sitting by the mess tent back at Mkangira, smoking her cigarettes and drinking her tea, and gazing with mild bafflement at the darkening river, winding down across that part of Africa where she has spent most of her life.

Over Brian's shoulder I watch Goa; he has gone down to wash, and now sits on his heels by the river's gleam like a driftwood stump. Each evening Goa comes to receive the instructions for the next day from Bwana Niki, and now he approaches and hunches down at a little distance, a mute, dark silhouette on the river bluff. Brian has not noticed him, but when I point, says, "Goa."

"This man here was always interested in the animals," Brian murmurs, as Goa comes and sits down on the ground nearby. "He really *cares*." In these days on foot safari Brian has spoken with affection and respect not only of Goa but of many Africans he has known and worked with in the bush, granting them status as companions, as real people he could like and trust. When I mention this, he frowns. "If you find someone you can work with, out in the bush," he says, "someone you can trust, you're bound to become friends."

In recent years, Goa Mwakangaru has married a Taita woman sent down to him by their families, and he has two children. Now he wishes to return to Kenya, although he knows that if he does so there is almost no chance the Tanzanian authorities will send him his Game Department pension. Squatting on his heels near Bwana Niki, as content as ourselves to gaze out on the river, Goa says quietly, "All the good work that we did here in the old days is being ruined. There is nothing for me here in the Selous. I am discouraged, and I would like to return to my own people in the Taita Hills."

During the night, hyenas draw near to vent their desolate opinions, and toward daybreak the lions are resounding. "Never heard them at all," the Warden grumps, sipping his tea in the gray-pink light of the dawn sky; he has slept badly on the narrow camp cots that in recent years have replaced the sturdy safari cots of other days. "Don't like to miss the lions in the night. Never get sick of that sound, no matter how often I hear it."

In the red sunrise, a pair of pied kingfishers cross the path of light on the shallow river to the palm fronds overhead, and mate in a brief flurry in the sun's rays. With the light, the ground hornbills are still, and doves and thrushes rush to fill the silence.

A cool wind out of the south; we head downstream. A herd of impala, the emblematic antelope of Africa, springs away over the green savanna, and as we pass, the great milky-lidded eagle owl eases out of a thick kigelia and flops softly a short distance to another tree, pursued by the harsh racket of a roller. Already tending north-northwest toward its confluence with the Luwegu, the river unwinds around broad sand bars and rock bends, and wherever it winds away toward the east, the man called Goa cuts across the bends, following the river plains, the hills, the open woods, and descending once again to the westward river.

NOTES

Chapter I

1. And the second largest in the world, after the Wood Buffalo Park in northern Alberta, which can claim scarcely twenty species of large mammals, as opposed to thirty-six in the Selous.
2. Formerly the Queen Elizabeth and the Murchison Falls National Parks.
3. See New York *Times*, 18 August 1979.
4. Conversation with Dr. Thomas Struhsaker, New York Zoological Society, 16 October 1979. See also Karl Van Orsdol, 'Slaughter of the Innocents', *Animal Kingdom*, Dec. – Jan. 1979.

Chapter II

1. Conversation with W. A. Rodgers, August 1979.
2. Helge Kjekhus, *Ecology Control and Economic Development in East African History*, London: Heinemann, 1977.
3. Margaret Lane, *Life with Ionides*, London: Hamish Hamilton, 1963.
4. W. A. Rodgers, 'The Sleeping Wilderness', *Africana*.
5. Alan Wykes, *Snake Man*, London: Hamish Hamilton, 1964.

Chapter III

1. Brian D. Nicholson, 'The African Elephant', *African Wildlife*, Vol. 8, Part IV, pp. 313–22 (1954).
2. Conversation with Dr. Thomas Struhsaker, 16 October 1979.
3. Conversation with Dr. David Western, October 1979.

Chapter IV

1. At the Mweka College of Wildlife Management, at Moshi.

Chapter VII

1. Nicholson, 'The African Elephant'.

Chapter IX

1. Wykes, *Snake Man*.
2. See R. M. Bell, 'The Maji-Maji Rebellion in Liwale District', *Tanganyika Notes and Records*, 1950.
3. K. Weule, 'Native Life in East Africa', London, 1909, quoted in Kjekshus, *Ecology Control and Economic Development in East African History*.
4. J. P. Moffett, *Handbook of Tanganyika*, 1958.

[*213*]

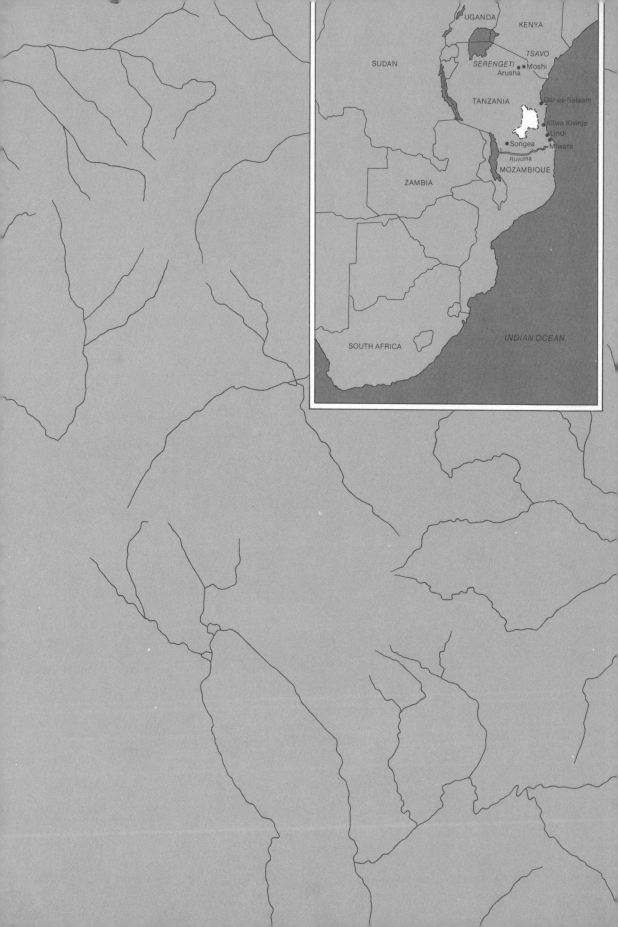